Harnessing the
Internet of Things (IoT)
for a
Hyper-Connected Smart World

Harnessing the
Internet of Things (IoT)
for a
Hyper-Connected Smart World

Edited by
Indu Bala, PhD
Kiran Ahuja, PhD

APPLE
ACADEMIC
PRESS

First edition published 2023

Apple Academic Press Inc.
1265 Goldenrod Circle, NE,
Palm Bay, FL 32905 USA
760 Laurentian Drive, Unit 19,
Burlington, ON L7N 0A4, CANADA

CRC Press
6000 Broken Sound Parkway NW,
Suite 300, Boca Raton, FL 33487-2742 USA
4 Park Square, Milton Park,
Abingdon, Oxon, OX14 4RN UK

© 2023 by Apple Academic Press, Inc.

Apple Academic Press exclusively co-publishes with CRC Press, an imprint of Taylor & Francis Group, LLC

Library and Archives Canada Cataloguing in Publication

Title: Harnessing the Internet of things (IoT) for a hyper-connected smart world / edited by Indu Bala, PhD,
    Kiran Ahuja, PhD.
Names: Indu Bala (Lecturer in electronics and electrical engineering), editor. | Ahuja, Kiran, 1981- editor.
Description: First edition. | Includes bibliographical references and index.
Identifiers: Canadiana (print) 20220222584 | Canadiana (ebook) 20220222614 | ISBN 9781774639740 (hardcover) |
    ISBN 9781774639757 (softcover) | ISBN 9781003277347 (ebook)
Subjects: LCSH: Internet of things—Industrial applications.
Classification: LCC TK5105.8857 .H37 2023 | DDC 004.67/8—dc23

Library of Congress Cataloging-in-Publication Data

....

CIP data on file with US Library of Congress

....

ISBN: 978-1-77463-974-0 (hbk)
ISBN: 978-1-77463-975-7 (pbk)
ISBN: 978-1-00327-734-7 (ebk)

About the Editors

Indu Bala, PhD

Associate Professor, School of Electronics and Electrical Engineering,
Lovely Professional University, Phagwara, Punjab, India

Indu Bala, PhD, is an Associate Professor in the School of Electronics & Electrical Engineering at Lovely Professional University, Phagwara, Punjab, India. Her research interests include cognitive radio networks, dynamic spectrum access, interference mitigation, and power allocation for cognitive radio networks. She received her BTech and MTech degrees in Electronics and Communication Engineering from Punjab Technical University, Jalandhar, India, and her PhD from Punjabi University, Patiala, India.

Kiran Ahuja, PhD

Assistant Professor, DAV Institute of Engineering and Technology,
Jalandhar, Punjab, India

Kiran Ahuja, PhD, is an Assistant Professor at the DAV Institute of Engineering and Technology, Jalandhar, India. She completed her post-doctoral fellowship at NIT, Jalandhar, under the N-PDF scheme of the Science and Engineering Research Board, India. She received her BTech and MTech degrees in Electronics and Communication Engineering from Punjab Technical University, Jalandhar, India, and her PhD in Electronics & Communication Engineering in wireless communication and networks field from Thapar University, Patiala, Punjab, India.

Contents

Contributors

Ayush Kumar Agrawal
National Institute of Technology Delhi, Delhi–110040, India, E-mail: ayush6295@gmail.com

Rohit Anand
G.B. Pant Engineering College, New Delhi, India, E-mail: roh_anand@rediffmail.com

P. Avirajamanjula
Professor, EEE, PREC, Thanjavur, Tamil Nadu, India

Manisha Bharti
Assistant Professor, National Institute of Technology Delhi, Delhi, India

Vinay Bhatia
Chandigarh Engineering College, Landran, Mohali, Punjab, India

Murthy Cherukuri
Department of Electrical and Electronics Engineering, National Institute of Science and Technology, Berhampur–761008, Odisha, India

Deepak Dadwal
Chandigarh Engineering College, Landran, Mohali, Punjab, India

P. Dananjayan
St. Peter's Institute of Higher Education and Research, Chennai, Tamil Nadu, India

J. Deepika
Department of Information Technology, Bannari Amman Institute of Technology, Tamil Nadu, India

Tanvika Garg
Research Scholar, National Institute of Technology Delhi, Delhi, India

Tulika Garg
Research Scholar, Punjab Engineering College, Chandigarh, India, E-mail: tulika.garg.bppc@gmail.com

Santamanyu Gujari
Sensor Specialist, KONE Elevator India Pvt. Ltd., Maharashtra–411001, India

Sapna Juneja
B.M. Institute of Engineering and Technology, Sonepat, Haryana, India

Harmandar Kaur
Department of Engineering and Technology, GNDU RC, Jalandhar–144007, Punjab, India, E-mail: harmandargndu@gmail.com

Simarpreet Kaur
Department of Electronics and Communication Engineering, BBSBEC, Fatehgarh Sahib, Punjab, India, E-mail: rabbyshabby6@gmail.com

B. Praveen Kumar
Bharat Institute of Engineering and Technology, Hyderabad, Telangana, India

Ch. Santhan Kumar
Lords Institute of Engineering and Technology, Himayath Sagar, Hyderabad, Telangana, India

K. Ashok Kumar
Matrusri Engineering College, Hyderabad, Telangana, India, E-mail: kashok483@gmail.com

Vankadara Sampath Kumar
Bharat Institute of Engineering and Technology, Hyderabad, Telangana, India,
E-mail: sampath.vankadara62@gmail.com

Palvinder Singh Mann
DAV Institute of Engineering and Technology, Jalandhar, Punjab–144008, India; I.K. Gujral Punjab
Technical University, Punjab–152004, India, E-mail address: psmaan@davietjal.org

Mani Raj Paul
CT Institute of Technology and Research, Maqsudan, Jalandhar, Punjab, India

H. D. Praveena
ECE, Sree Vidyanikethan Engineering College, Tirupati, Andhra Pradesh, India

C. Rajan
Department of Information Technology, K. S. Rangasamy College of Technology, Tamil Nadu, India

Vanga Karunakr Reddy
Matrusri Engineering College, Hyderabad, Telangana, India

Amiya Ranjan Senapati
Project Engineer, Wipro Technologies, Karnataka–560035, India

T. Senthil
Department of Electronics and Communication Engineering, Bannari Amman Institute of Technology,
Tamil Nadu, India, E-mail: t.senthilece@gmail.com

Anshu Sharma
CT Institute of Technology and Research, Maqsudan, Jalandhar, Punjab, India

Anurag Sharma
GNA University, Phagwara, Punjab, India, E-mail: er.anurags@gmail.com

Himanshu Sharma
J.B. Institute of Engineering and Technology, Hyderabad, Telangana, India

Nidhi Sindhwani
Amity University, Noida, Uttar Pradesh, India

Satvir Singh
DAV Institute of Engineering and Technology, Jalandhar–144008, Punjab, India;
I.K. Gujral Punjab Technical University, Punjab–152004, India

Ankur Singhal
Chandigarh Engineering College, Landran, Mohali, Punjab, India

Tarun Singhal
Chandigarh Engineering College, Landran, Mohali, Punjab, India, E-mail: tarun.sgl@gmail.com

Kunjabihari Swain
Department of Electrical and Electronics Engineering, National Institute of Science and Technology, Berhampur–761008, Odisha, India, E-mail: kunja.swain@gmail.com

G. S. Vinoth
QUEST Global, Technopark, Thiruvananthapuram, Kerala, India

N. Vithyalakshmi
ECE, Sree Vidyanikethan Engineering College, Tirupati, Andhra Pradesh, India, E-mail: vidhyavinoth@gmail.com

Abbreviations

A/D	analog-to-digital
ABC	artificial bee colony
ACO	ant colony optimization
AG-AG	aboveground to aboveground channel
AI	artificial intelligence
AR	augmented reality
BA	bus arbiter
BER	bit error rate
BS	base station
CDMA	code division multiple access
CH	cluster head
CIoT	cognitive Internet of Things
CPU	central processing units
CR	compression ratio
CWSI	crop water stress index
D/A	digital-to-analog
D2D	device-to-device
DE	differential evolution
DEECIC	distributed energy-efficient clustering with improved coverage
DRRA	distributed round-robin arbiter
DSA	dynamic spectrum access
EA	evolutionary algorithm
ECG	electrocardiography
EDA	estimation of distribution algorithms
EEMC	energy-efficient multi-level clustering
EESA	energy efficient spectrum access
EH	energy harvesting
EHT	energy harvesting technique
EMF	electromotive force
EMI	electromagnetic interference
ERP	evolutionary routing protocol
FAO	Food and Agriculture Organization

FPGA	field-programmable gate arrays
GA	genetic algorithm
GALS	globally asynchronous locally synchronous
GIS	geographic information systems
GPIO	general purpose input/output
GPR	ground penetrating radar
GPS	global positioning systems
GSM	global system for mobile communication
H2M	human to machine
HAS	harmony search algorithm
HEED	hybrid energy-efficient distributed clustering
HOL	head of blocking
iABC	improved artificial bee colony
ICT	information and communication technology
IoT	Internet of Things
IoUT	internet of underground things
IP	intellectual property
ISI	inter symbol interference
ITRS	international technology roadmap for semiconductor
KNN	K-nearest neighbor
LAN	local area network
LEACH	low-energy adaptive clustering hierarchy
LiDAR	light detection and ranging
LR	logistic regression
LUT	lookup table
M2M	machine-to-machine
MAC	medium access control
MBC	mobility-based clustering
MCN	maximum cycle number
MI	magnetic induction
ML	machine learning
MQTT	message queuing telemetry transport
MRP	multi-path routing protocol
MRS	microwave remote sensing
NDVI	normalized difference vegetation index
NFC	near field communication
NI	network interface
NLP	nonlinear programming

NoC	network on chip
NOMA	non-orthogonal multiple access
OCI	overloaded CDMA interconnect
OCR	optical character recognition
OFDM	orthogonal frequency division multiplexing
PEACH	power-efficient and adaptive clustering hierarchy
PEs	processing elements
PN	population number
PRD	percentage root mean squared difference
PRRA	parallel round-robin arbiter
PSO	particle swarm optimization
PTS	parallel to serial
QoS	quality of service
RDBMS	relational database management system
RRA	round-robin arbiter
SA	switch arbiter
SHM	structural health monitoring
SoC	system on chip
SQL	structured query language
SVM	support vector machine
UAV	unmanned aerial vehicles
UG-AG	underground to aboveground channel
VCT	virtual-cut through
VISA	virtual instrument software architecture
VR	virtual reality
VRI	variable rate irrigation
VRT	variable rate technology
WH	worm-hole
WSNs	wireless sensor networks
WUCNs	wireless underground communication networks
WUSN	wireless underground sensor network

Acknowledgments

The editors are grateful to Apple Academic Press for permitting them to edit this book, *Harnessing the Internet of Things (IoT) for a Hyper-Connected Smart World;* the support, suggestions, and encouragement offered are praiseworthy.

We are thankful to the authorities of Lovely Professional University, Phagwara, Punjab, and DAV Institute of Engineering and Technology, Jalandhar, Punjab, for providing a congenial atmosphere and moral support to carry out our work in a smooth manner.

Also, we are thankful to all the authors for their insightful contributions and the reviewers for their timely support and constructive suggestions, which have improved the quality of the chapters substantially.

Also, many colleagues and friends have helped in some way or another. We wish to thank them all.

Preface

For the past few years, the term *Internet of Things* (IoT) has grabbed attention to provide a global infrastructure of networked physical objects called "things." These physical objects are embedded with sensors, actuators, software to achieve certain goals or services by exchanging information with other connected objects over the internet. IoT has revolutionized the world by embedding intelligence and any-time, anywhere connectivity in real-time objects to transform them into smart objects. The basic notion of IoT is to unite every object of the world under one common infrastructure; in such a manner that humans not only can control those objects, but to provide regular and timely updates on the current status. With this background, this book, *Harnessing the Internet of Things (IoT) for a Hyper-Connected Smart World*, is an attempt to explore the concepts and applications related to the Internet of Things with the vision to identify and address existing challenges. The book will appeal to students, practitioners, and researchers working in the field of IoT to integrate IoT with other technologies to develop comprehensive solutions to real-life problems.

This book is organized into 13 chapters. A brief description of each chapter is as follows.

Network on Chip (NoC) is the new paradigm for System on Chip (SoC) integration to maintain high performance for IoT applications. The router structure is one of the critical design issues for efficient NoC communication architecture. To address this issue, the authors in Chapter 1 have presented the router design to attain high speed. The scheduling algorithm and structure of the crossbar are improvised for high-speed IoT applications.

IoT systems are the driving force behind the significant changes in the traditional educational system by making it more interactive between a student and a teacher. IoT has the tremendous potential to incorporate data-driven decision-making into every aspect of human activity. The network of sensors and actuators in an IoT system is blurring the lines between the physical and digital worlds. Despite significant technological progress over the past few years, the adoption of smart classroom technology has been rather late in leveraging change in teaching methods. In Chapter 2, the authors have discussed the broad spectrum of IoT applications in the educational industry along with the challenges associated with them.

As the global populace increases, it is imperative to work on developing methods to enhance crop productivity with lesser use of natural resources and assets in a progressive way. It is believed that digital agriculture can play a major role in solving problems of the agriculture sector. Advanced Farming techniques using digital innovations portray the development in horticulture and agribusiness from precise cultivation to associated, information-based ranch creation frameworks. Advanced farming utilizes precision farming innovation along with smart systems and innovating techniques to transform the horticulture segment and bring huge advantages for crop growers. Digital farming is the science in which information technology-enabled ecology helps in the growth and development of services on time to make cultivating beneficial and feasible while producing quality food for all at a nominal rate. To stress the dire need for digital irrigation in society, Chapter 3 deals with diverse issues concerning agriculture and challenges along with solutions with the help of the latest technical smart tools to uplift the farming community.

IoT applications are becoming increasingly popular in modern society due to their faster adoption in our daily lives. The health care industry is no different, and every day new products are coming into the market for patient monitoring and care. In Chapter 4, the authors have presented an IoT-based smart jacket design comprised of smart sensors to monitor heart rate, sugar level, blood pressure, fever, and stress level. A web-based application is designed to monitor sensor data in real-time by the doctor(s). This smart jacket can be GPS enabled, which automatically updates the location of the patient whenever the person goes from one point to another and is accessed live by the doctor.

Various technologies are integrating to embed intelligence into IoT systems. Cognitive Internet of Things (CIoT) refers to the use of various technologies related to cognitive computing for self-organized interconnected machines to enhance the efficiency of the sensor-based IoT system. In Chapter 5, a brief survey related to CIoT is presented, followed by a detailed discussion on various functionalities of the cognitive IoT system. Various state-of-the-art techniques like optimization schemes, blockchain, machine learning (ML), Orthogonal Frequency Division Multiplexing (OFDM), etc., are discussed in context to the cognitive IoT. Moreover, various applications of CIoT in sensor networks, health monitoring, energy harvesting, neural networks, data analytics, 5G networks, and industries are also discussed. At the end of the chapter, research challenges are discussed with proposed solutions.

Wireless Underground Sensor Networks (WUSNs) have a wide range of applications in military, underground sensing, testing soil traits and moisture content, pollution control, location detection, security, and detection of natural calamities. Despite their huge prospective, the growth of WUSNs has to cope with several research challenges due to the complex concealed environment. In Chapter 6, the author has discussed key characteristics and classification of WUSNs, and the research challenges associated with them. Potential application areas for each layer of the protocol hierarchy are also presented with the future scope of improvement at the end of the chapter.

Clustering is a well-known optimization problem on the Internet of Underground Things (IoUT) networks comprised of multiple sensors. Some Computational Intelligence (CI)-based metaheuristic approaches are proven to be competitive over other conventional analytical techniques to solve optimization problems in Underground Wireless Sensor Networks (UWSNs). In Chapter 7, the authors have presented an improved Artificial Bee Colony (iABC) metaheuristic algorithm to yield the optimal solution to the problem. The proposed iABC2 protocol is proposed to obtain optimal cluster heads (CHs) in UWSNs. At the chapter's end, simulation results are presented to manifest that iABC2 outperforms other well-known cluster-based protocols based on energy consumption, network lifetime, and the end-to-end delay.

With enormous advancement in the area of IoT, innovative and dynamic infrastructures are being deployed in everyday life. On realizing the need for IoT systems for process monitoring, quality control, the safety level of employees, and improving the working environment in a hazardous condition, IoT-based real-time monitoring solutions are in demand. In Chapter 8, the authors have presented an IoT-based prototype to mimic the effective monitoring and controlling of industrial pollutants and disposal of industrial waste, using Arduino ATmega328 high-performance microcontroller.

WUSNs are used to monitor various conditions such as toxic substances and soil properties in agriculture. Unlike the present underground sensors that are connected to the ground via wire, WUSNs are completely buried underground with no such wired connections. This is the advantage of WUSNs over existing underground sensors. Through dense substances such as soil or rock, wireless communication is quite difficult; even saving energy is also a major challenge for WUSNs. Thus, this requires redesigning communication protocols. The authors have addressed various design challenges and applications of WUSNs in Chapter 9. At the end of the chapter, different aspects of each layer of the communication protocol stack are also discussed.

Most of the IoT-based products/applications are battery-operated. Since the batteries have a limited life span; they need to be replaced after some time. The task of battery replacement becomes more tedious when IoT-enabled devices are deployed randomly in a remote area. Energy harvesting is one of the best cost-effective solutions to provide batteries for an unlimited time, especially in situations where it is difficult to replace them. In Chapter 10, the authors have presented a comprehensive review of various energy harvesting techniques to generate electrical power from non-conventional power sources with their merits and demerits. Moreover, a hybrid energy harvesting method is also proposed to overcome the drawbacks of standalone renewable energy sources.

In Chapter 11, IoT-based Peltier air conditioner design is proposed by the authors. The novel design utilizes the Peltier Module to overcome all the limitations of the existing HVAC framework. The authors have reviewed the existing cooler designs, advantages of Peltier coolers, and their applications over conventional cooling gadgets. The proposed IoT-based design using thermoelectric Peltier model performance is evaluated and compared with the state of art cooler design.

IoUT agricultural software permits the gathering of relevant data by ranchers and farmers. Wide landowners and small farmers need to realize the IoUT market's opportunity for agriculture by implementing smart technology to improve their outputs' productivity and sustainability. With the increasingly increasing population, demand can be fulfilled effectively if the ranchers, as well as the small farmers, are effectively introducing agricultural IoUT solutions. In Chapter 12, the authors have addressed various agricultural issues being faced by the farmers, and IoUT-based solutions have been proposed to address some of them to bridge the gap between production and yielding quality and quantity.

IoT has bought a new direction in the domain of agriculture from an innovative research perspective. It has wide application in various areas in agriculture, and since it is in its initial stages, there is a lot of room for experiment. In Chapter 13, the various potential IoT sensing devices are reviewed that are crucial to ensure an improved and smart farming experience. The various sensor categories, their working dynamics, and the specific suitable areas are focused on and analyzed comprehensively. The sensor-enabled IoT is indispensable for revolutionizing farm and agriculture practices. Various standalone sensors and sensor systems are presented, along with the discussion on the role of various benchmark commercial sensors available in the market that provide intelligent services to the farmer. The chapter helps to understand the innovations in sensors so far, and additionally, a future roadmap is also presented.

To draw the various issues in IoT, challenges faced, and existing solutions so far, the chapters of this volume have been meticulously selected and studied by knowledgeable reviewers. It is the wish of the editors of this volume that this effort of theirs in accumulating so many challenges faced in the field of IoT, which have been focused on in the various chapters of this volume, will be helpful for future research in this field. Specifically, the real-world problems and different application areas presented will attract the attention of the researchers in the field and provide them with valuable input.

*—**Editors***

CHAPTER 1

Design of NoC-Based High-Speed Processor to Suit IoT for a Hyper-Connected Smart World

K. ASHOK KUMAR,[1] VANGA KARUNAKR REDDY,[1] and P. DANANJAYAN[2]

[1]*Matrusri Engineering College, Hyderabad, Telangana, India, E-mail: kashok483@gmail.com (K. A. Kumar)*

[2]*St. Peter's Institute of Higher Education and Research, Chennai, Tamil Nadu, India*

1.1 INTRODUCTION

As scaling of technology decreases and reduces the feature size leads to an increase in the number of Intellectual Property (IP) cores and functional blocks in a System on Chip (SoC). According to International Technology Roadmap for Semiconductor (ITRS)-2011 report, the feature size will further reduce to 14 nm by 2020, and also popular companies like Intel Corporation, NVIDIA Corporation are forecasted that the feature size will come to 11 nm by 2020. The SoC has a huge number of on-chip components that are connected with bus-based communication. When a system with parallel processing, the bus-based communication is ineffective and does not satisfy the performance requirements; thereby, the communication is a major bottleneck for the applications of SoC. A Network on Chip (NoC) is the new paradigm to handle overheads of intercommunication among Processing Elements (PEs) [4]. The NoC is mainly composed of

Harnessing the Internet of Things (IoT) for Hyper-Connected Smart World.
Indu Bala and Kiran Ahuja (Eds.)

three elements, namely routers, network interface (NI), and interconnect wires, as shown in Figure 1.1. The router/switch is the heart of NoC communication architecture because it controls the entire communication in NoC-based SoC [13]. The function of NI is to packetize and de-packetize of data packets from/to PE. The bi-directional interconnecting channel is responsible for transferring the data between the Input-Output (I/O) ports. The router is mainly realized with a number of I/O ports, arbiter modules, and crossbar switches.

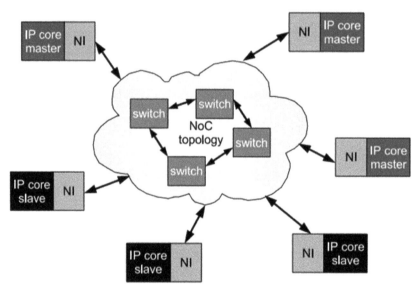

FIGURE 1.1   Structure of NoC-based communication.

A typical switch consists of a total of five I/O ports, including four directional ports and one local port for efficient data transfer between the source PE and destination PE. Each port has limited memory to store data packets temporarily; thereby, th e deadlock error is avoided efficiently. The arbiter solves the arbitration among ports and the port transfers data packets which gets grant signal from the arbiter. Different scheduling algorithms are proposed to increase the operation speed of the arbiter module. The crossbar switch efficiently transfers data packets from selected input to output port hence the design of multiple port supported crossbar switch is critically important for NoC router [3]. The different structures of crossbar switch are proposed to increase the data transfers speed of NoC-based SoC. Along these, the other design constraints are also affects the performance of NoC communication architecture namely, topology, switching technique and routing algorithm. Figure 1.2 shows the structure of mesh-based NoC communication architecture. The inner layer of NoC routers is connected with four neighbor routers and their corresponding PE. The structure of 2-dimensional (2-D) mesh topology and routing of data transfer are clearly shown in Figure 1.2. The various topologies are applied and tested their performance for NoC communication architecture namely mesh, torus, ring, tree, butterfly, and star. The different routing algorithms are proposed to complete the data transfer from source PE to destination PE with less delay because several intermediate routers involved for completing the data transaction. The data switching technique gives how data packet transfers from one router to another router. The two different data switching are commonly used for NoC communication namely packet switching and circuit switching.

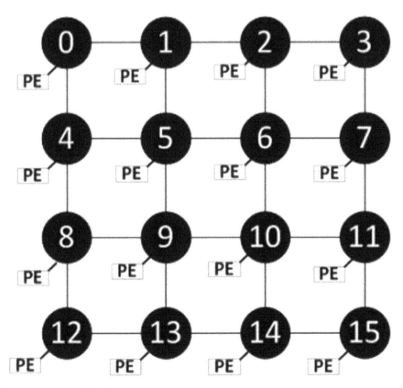

FIGURE 1.2   4×4 Mesh-based NoC communication architecture.

The modifying of all design constraints in NoC fabric are impractical. In this chapter, the design constraints of the NoC router are advanced and improved that is improvement in scheduling algorithm, designing of crossbar, and routing algorithm. The performance of advanced NoC fabric is evaluated and compared in terms of occupied area, latency, power consumption and throughput with recent literature. This remaining chapter as follows; Section 1.2 presents a state of art for existing literature and Section 1.3 describes an overview of NoC architecture. Section 1.4 provides the main contributions of the chapter and Section 1.5 propounds on the implementation results, and finally, Section 1.6 concludes the chapter.

1.2 BACKGROUND

The state of art for NoC fabric with existing literature to design the NoC router thereby obtains high performance for SoC. Bus Arbiter (BA) is proposed based on Round Robin Arbiter (RRA) to handle four input requests [10]. The input requests of BA are handled by round-robin arbiter generator (RAG) tool and produce the grant signal. The concept of token-ring network is used to implement BA. The control of token allows enabling the priority logic block that is varied the priorities of input requests. The output of 4-bit ring counter is given to enable signal of priority logic block and also generated the tokens. The Switch Arbiter (SA) of 2×2 and 4×4 is implemented with the help of BA thereby RAG tool generates MxM SA to enhance the performance of arbiter module in NoC router. However, the area and delay are increase linearly with increase of M. To reduce the arbitration delay and increase the speed of arbiter, a Parallel Round Robin Arbiter (PRRA) is proposed based on simple binary search algorithm and also proposed improved PRRA (IPRRA) to reduce the timing of PRRA [14]. The process of PRRA consists of two traces that are up-trace and down-trace. The function of up-trace is collecting input requests and their priority information, whereas the function of down-trace is making decision depend on the data provided by up-trace. The IPRRA is proposed to resolve arbitration because of PRRA consumes more gate delay. By overlapping up and down traces, the IPRRA is provided improved performance than PRRA and SA. To increase clock frequency, Parallel Pseudo Round Robin (P2R2) is proposed based on the idea of multiple groups executing in parallel [2].

The P2R2 divides the total input request into a number of equal groups thereby all groups execute parallelly. In order to realize parallelism, the number of bits are asserted in the priority vector of P2R2 arbiter, and the asserted bits are responsible for propagating priority bits to the number of groups. Because of parallelism, the P2R2 arbiter improves the fairness, clock period while the area and power overheads are increase with increase of a number of groups. To reduce area and power overhead, a distributed round-robin arbiter (DRRA) is proposed [9]. DRRA is distributed the incoming requests cyclically to choose single request from multiple input request thereby the efficiency of the DRRA is improved then typical RRA. However, the efficiency is reduced when the number of input requests are more. To obtain high performance of arbiter in case more of input requests, a modified distributed round-robin arbiter is proposed in this book chapter.

The structure of the crossbar switch is realized with Code Division Multiple Access (CDMA) recently. The novel Walsh-based CDMA is proposed for crossbar switch with star topology of NoC fabric to handle a large number of input ports [7]. The local switches are communicated via the central switch of star network topology to get high throughput and low delay. To support synchronous and asynchronous communication, a Globally Asynchronous Locally Synchronous (GALS)-based CDMA proposed for crossbar switch of NoC router [12]. The issues by applying CDMA to the NoC router are discussed and also the importance of spreading codes is identified. The Walsh-based CDMA with GALS-based NoC is proposed to reduce data packet latency. Still, the area and power requirements are increased because of more transition activities are needed to complete the data transfer. To reduce data transmission latency, a novel architecture is proposed with sharing of resources among the senders [5]. By using dynamic assignment of spreading codes, the system with a large number of users reduces the encoding and decoding circuitry, and also their delays are reduced drastically. Still, the network latency increases because of dynamic assignment of spreading codes. A Standard Basis (SB)-based codes are used in place of Walsh-based codes for CDMA encoding and decoding operations of NoC router [11]. The proposed method showed better area utilization and latency than Walsh-based CDMA. To obtain high bandwidth, an Overloaded CDMA Interconnect (OCI) is proposed with an increase of usable spreading codes [1]. The serial and parallel OCI are presented to enhance system capacity and also achieve low area and delay requirements. The parallel OCI is showed 100% higher bandwidth than the conventional CDMA crossbar switch. However, by using orthogonal and non-orthogonal spreading codes increases the data packet latency and power consumption. To achieve high throughput and bandwidth, a parallel Overloaded CDMA with orthogonal gold codes is proposed in this book chapter. By the help of parallel encoding and decoding, the parallel OCDMA is showed better results than the existing CDMA structures of NoC router.

1.3   OVERVIEW OF NOC ARCHITECTURE

As mentioned earlier, the performance of NoC communication architecture depends on various design constraints. Among the various constraints, network topology, data switching techniques, routing algorithm, and buffering strategy are critically important for NoC communication architecture. The structure of NoC router directly affects overall performance of SoC hence the design router is majorly important.

1.3.1   NETWORK TOPOLOGY

The network topology is defined as the structure of nodes how arranged to form a network. The different types of topologies are implemented for on-chip interconnects.

Figure 1.3 depicts the various topologies are used for NoC-based SoC. The entire nodes of the network are connected in the form of a circle called ring topology, as shown in Figure 1.3(a). The nodes are connected either unidirectional or bi-directional to get high performance. Still, if any node fails, the remaining network will fail because of the loss of connection between source and destination. The mesh topology which is connected neighbor nodes in 2-dimensional (2-D) fashion is shown in Figure 1.3(b). As a number of available interconnection links is more in mesh topology, the data packet experiences less traffic between intermediate nodes. The degree of edge nodes is less when compared with the degree of centered nodes in mesh topology. To increase the degree of edge nodes, torus topology is introduced by extending the mesh topology. A wrap-around connections is created for connecting of edge nodes; thereby, the degree of center and edge nodes are almost equal. Though, the interconnection wires are increased because of wrap-around connection. To get compatibility of nodes in a network, the structure of torus is advanced and formed like folded manner; thereby, the folded torus is used for on-chip network to improve the performance of torus topology. The structure of folded torus topology is shown in Figure 1.3(d). The specific application needs a tree-based topology and the structure of fat-tree topology is shown in Figure 1.3(e). In fat-tree-based topology, one root node controls the connected leaf nodes hence each layer of nodes connected top layer and bottom layer of nodes.

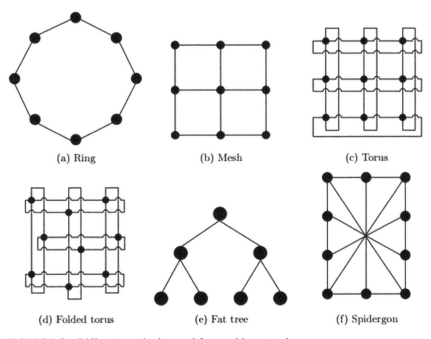

| (a) Ring | (b) Mesh | (c) Torus |
| (d) Folded torus | (e) Fat tree | (f) Spidergon |

FIGURE 1.3   Different topologies used for on-chip networks.

The Spidergon topology also compatible for NoC-based SoC and the structure of Spidergon is shown in Figure 1.3(f). The entire nodes are connected with a center node; hence, the connections of different nodes are connected with a center point. The other different structures of topologies are used for on-chip interconnects, namely, star-based topology, octagon topology, butterfly topology, and irregular topologies for application-based topologies.

1.3.2   *ROUTING ALGORITHM*

Routing algorithm defined that the way data how transferred from the source to destination through intermediate routers. Depending on the topology used for NoC, the specific routing algorithm is utilized to improve the speed of data transfer. The various routing algorithms are proposed for NoC to improve the performance of the NoC-based system. The routing algorithms are mainly classified into three categories:

- Routing decision;
- Routing path length; and
- Routing path determination.

The first category of routing algorithms is based on routing decisions. The source routing and distributed routing are examples of routing decisions. The entire routing path of data from source to destination is calculated before data transmission. The routing path is fixed and not to change during the data transmission. The fixed path for data enhances the speed of data transfer; thereby, data rate increases and latency of data packet reduces. Still, the data packet experiences high traffic because of the fixed data path from source to destination.

The secondary category of routing algorithms is based on the routing path length. The minimal and non-minimal routing algorithms are examples of path length. The minimal routing algorithm transfers the data based on the shortest path between the source and destination. Each router determines the available short path towards the destination during data transmission. The non-minimal routing algorithm determines the data route in 2-D; thereby, increasing the flexibility and avoiding the live-lock error. The routing algorithms that are based on routing path length have the possibility to occur network errors like deadlock, live-lock, and starvation.

The third category of routing algorithms is based on path determination. The deterministic and adaptive routing algorithms are examples of path determination. In a deterministic routing algorithm, the entire routing path is specified from source to destination completely. The deterministic routing algorithms transfers the data to the destination towards one direction; thereafter, data transfer to another direction of the destination router. The adaptive routing algorithm transfers the data in any direction dynamically.

The specific routing algorithm used to data transfers based on the application of NoC-based SoC. each category has advantages and disadvantages in terms of network performance.

1.3.3 SWITCHING

The data switching technique determines how the data packet routes in the network. The two types of switching techniques are used for on-chip interconnects namely:

- Circuit switching; and
- Packet switching.

In the circuit switching, a dedicated path is created between the source and destination for efficient data transfer. The dedicated circuit set up establishes thereby the data transfers started from source to the destination. The modern telecommunication network is the best example for circuit switching. The interconnection link is reserved until data transfer is completed from source to destination. The data transfer speed increases because of the reserved path between routers. Still, the circuit switching has a demerit in terms of circuit delay and low saturation point.

The packet switching techniques popularly used for on-chip interconnect networks. The packet switching transfers the data packets from router to router without creating reserved link; thereby, the data packet delay reduces. The packet switching is again classified into three categories that are Store and Forward (SF), Virtual-Cut Through (VCT), and Worm-Hole (WH) switching. The SF switching technique stores data before sending to the neighbor router; hence, each router needs the buffer memory for storing the entire data packet. The VCT switching technique transfers the data by dividing the data packet into various parts virtually. The data entire packet is stored virtually in entire routers; thereby, the requirement of buffer memory reduces [20]. Still, data packet latency is not reduced because data transfer starts to neighbor once the entire packet is stores in the current router. To reduce memory requirement, the data packet divides the number of parts, as shown in Figure 1.4. The data packet is divided into flow control unit (FLIT), and each flit is divided into physical control units (Phits). In WH switching, the data transfers by using low memory unit (Phit). The Phits are transferred through the intermediate routers with less memory from source to destination; thereby, the memory requirement and data packet delay are reducing. Still, the head of blocking (HOL) error occurs in the WH switching technique.

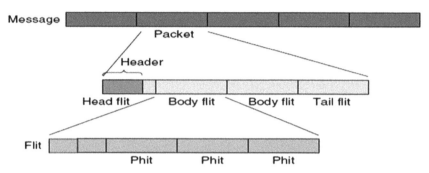

FIGURE 1.4   Data packet format.

1.3.4   BUFFERING STRATEGY

The buffer is used to store the data packets in intermediate routers when data transmission from source to destination; thereby, the deadlock error is efficiently avoided. Though, the size of the port increases because of requires memory for storing the entire data packet; hence, the area utilization of the entire system increases [8]. To reduce area utilization, the data packet is divided into various parts that are flits and Phits.

1.3.5   NETWORK INTERFACE

The function of network interface (NI) is connecting PE to the router to transmission of data packets. The NI converts and sends the data packets into suitable format to the router thereby each router needs NI to connect with the PE.

1.4   MAIN CONTRIBUTIONS

In this book chapter, the major contribution of research is advancing scheduling algorithm and proposing improved design for crossbar switch thereby improving the performance of NoC router and also NoC-based SoC. To enhance performance, an extra PE is added NoC router where typical NoC router is connected with single PE. The PE clustering improves the speed of data transfer without involving router thereby the requirement of more number of routers is reduced.

1.4.1 PARALLEL DISTRIBUTED ROUND ROBIN ARBITER (PDRRA)

The scheduling algorithm resolves the arbitration among the ports in NoC router. RRA is commonly used for arbiter module in NoC router. The different structures are proposed to improve the speed of scheduling algorithm. In this chapter, the classical RRA is modified thereby the speed arbiter module in NoC router improved and maintained same performance even in case of large number of ports for router. The structure of parallel distributed round-robin arbiter (PDRRA) divides the total number of input requests (N) by half that is, each block is equal number of input requests ($N/2$) and each block divided by half again. This division process is repetitive till each sub-block gets size of 2 input requests. Hence, the total number of input requests (N) are divided into various sub-blocks [$(log_2N) -1$]as shown in Figure 1.5, and the grant signal is selected from these sub-blocks with selector circuit that is designed with exclusive-or gate (EX-OR).

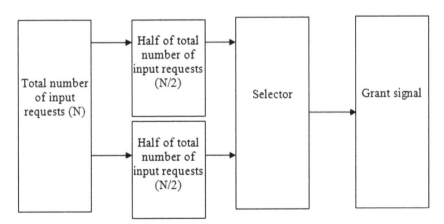

FIGURE 1.5   Structure of PDRRA.

Structure of PDRRA is modified from DRRA [9] to improve the speed and reduce the area utilization. The delay of DRRA is high when the number of input requests is more hence structure of DRRA is modified that maintain high performance even when more number of requests. The PDRRA is composed of a number of 2×1 arbiters which is made up of counter and multiplexer modules.

An entire 2×1 arbiters are executed parallelly, and the responses are given to the EX-OR-gate; thereby, the grant signal is selected from a number of input requests. This parallel execution increases the speed of arbiter for resolving the arbitration among input ports. The structure of 2×1 designed with combination of 1-bit counter and 2×1 multiplexer as shown in Figure 1.6. The output signal of multiplexer is given to input of counter module.

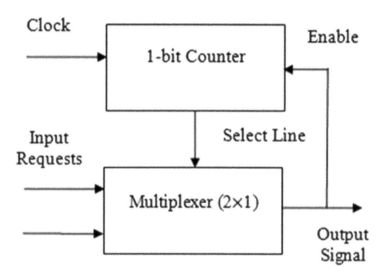

FIGURE 1.6   Structure of 2×1 arbiter.

The output signal of multiplexer initiates the counter module; thereby, the counter provides clock signals to the multiplexer with active low or active high signals. Depending on the counter signal, the multiplexer selects any one of the input requests and transfers to the output signal to the EX-OR-gate. These 2×1 arbiters send the output signals to the EX-OR-gate; thereby, the high performance is maintained even when a number of input requests are large. However, the power consumption of PDRRA is increased when compared with DRRA [9].

1.4.2 PARALLEL OVERLOADED CDMA

The function of the crossbar is to transfer the data packet from the input port to the output port. The structure of the crossbar switch is improved and advanced to get high speed of data transfer in NoC communication architecture. The typical crossbar switch is composed with a combination of multiplexer-demultiplexer circuit for transferring data packets when number ports are less. The CDMA is popularly used for function of crossbar switch of NoC communication architecture recently. The CDMA is a multiple access technique where several transmitters transfer the data simultaneously through a single channel which is perfectly suitable for NoC-based SoC. The spreading code is important for CDMA to avoid Inter Symbol Interference (ISI) and also Multiple Access Interference (MAI). Depending on the cross-correlation, the spreading code is selected for efficient data transfer. The spreading codes are classified into two categories that are orthogonal and non-orthogonal codes. The cross-correlation of orthogonal codes is zero whereas non-orthogonal codes not zero thereby the orthogonal codes are preferred for CDMA operation. A Walsh-based code are commonly used for orthogonal code. However, the code utilization of Walsh-based codes is less compared with non-orthogonal codes hence the research gap is identified. The structure of the crossbar switch improved by parallel overloaded CDMA with orthogonal gold codes in this chapter that is compatible for large number of inputs of CDMA. The orthogonal gold code provides larger code utilization than Walsh code with maintaining equal Bit Error Rate (BER). The generation of orthogonal gold codes by appending '0' for properly selected m-sequences. The overloaded CDMA [1] is modified and improved to get high performance of NoC fabric. Figure 1.7 depicts the parallel encoding circuit for OCDMA crossbar switch with orthogonal gold code. Each bit of spreading code is added arithmetically with data bit; thereby, the sum is got for each bit separately. The sum of each bit is encoded separately to the decoder section hence the data packets are converted into Parallel to Serial (PTS) before encoding and again data convert into Serial to Parallel (PTS) after decode circuit of CDMA crossbar switch. These converters are composed at NI of each router to efficient data packet transfer from the input port to the output port.

In the decoding section, the encoded sum bits are received individually; thereby, the original data is reconstructed with the help of orthogonal gold codes and shown in Figure 1.8. The orthogonal gold codes are multiplexed to encoded data at the decoder section again. Based on spreading code, the encoded data is divided and stored in either positive accumulator register or negative accumulator register. The comparison module selects the higher register between the values of positive and negative accumulator register. The original data reconstructed is either '*0*' when positive accumulative register higher than negative or '*1*' when positive accumulative register is lesser than negative. This process is the same as decoding of classical CDMA operation still, the decoding process is repeated for each encoded bit individually and executed these decoding process parallelly.

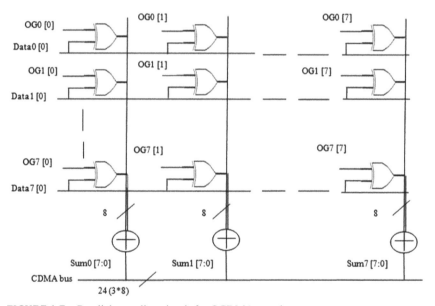

FIGURE 1.7   Parallel encoding circuit for OCDMA crossbar.

1.5   IMPLEMENTATION

The implementation of modified NoC router simulated Xilinx 14.7 software and 3×3 mesh-based NoC communication architecture simulated at Riviera-pro for windows. The performance metrics of area utilization, latency, power consumption and throughput are measured for evaluation of performance of NoC fabric and also compared with recent work. The simulation parameters considered for evaluation of improved NoC fabric are shown in Table 1.1. The overall performance of modified NoC router is evaluated by determining of performance of individual components thereafter combining the improved performance of each component. The performance of modified arbiter and parallel OCDMA crossbar switch are measured separately and also compared with recent literature. The area utilization is measured in terms of occupied slices, look up table (LUT)-flip-flop (FF) pairs, input-output buffers (IOB). The latency is measured in terms of delay (ns), Maximum clock frequency (MHz) and the total power consumption is measured combination of static and dynamic power consumption (mW). The throughput (bps) of mesh-based NoC is also measured for improved NoC fabric.

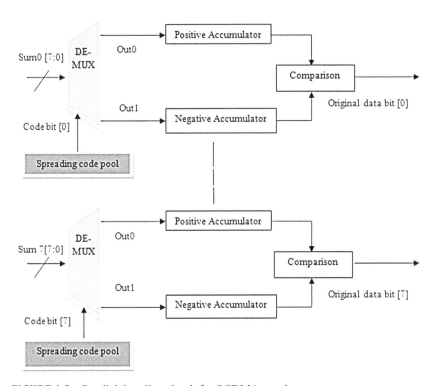

FIGURE 1.8 Parallel decoding circuit for OCDMA crossbar.

Table 1.2 is clearly shown that the performance of MDRRA is better than DRRA. Because of MDRRA is dividing of total input requests into a number of sub-blocks and executing in parallel, the performance of MDRRA is better than DRRA. The area utilization and power consumption of MDRRA is increased than DRRA because of dividing of total input requests into many sub-blocks. However, this increment in MDRRA is small and speed of servicing input requests constantly maintained even when the number of requests are increased.

TABLE 1.1   Simulation Environment

Simulation Parameter	Values
Topology	2-D mesh
Arbiter	Modified distributed round-robin
Switching	Store and forward
Crossbar switch	Parallel overloaded CDMA
Routing algorithm	Minimal adaptive
Data packet length	8 bit
Buffer	Yes
Simulator	Riviera-pro
Traffic scenario	Uniform random, transpose
Traffic distribution	Poisson
Virtual channels	No

TABLE 1.2   Performance Comparison of DRRA and Modified DRRA

Ports	DRRA [9]					PDRRA				
	Occupied Slices	LUT-FF Pairs	IOBs	Delay (ns)	Power Consumption (mW)	Occupied Slices	LUT-FF pairs	IOBs	Delay (ns)	Power Consumption (mW)
5×5	3	7	12	1.11	3.53	3	3	15	0.74	3.61
6×6	4	7	14	1.11	3.57	3	3	17	0.74	3.79
7×7	5	8	16	1.12	3.87	4	4	20	0.74	5.32
8×8	5	8	18	1.13	4.03	4	4	22	0.74	5.54
9×9	6	10	20	1.142	4.18	3	5	25	0.74	6.29
10×10	5	10	22	1.147	4.35	5	5	27	0.74	7.56
16×16	8	13	34	1.18	7.38	8	8	42	0.74	11.52
32×32	16	23	58	1.20	14.44	16	16	82	0.74	22.72

Figure 1.9 depicts the comparison of modified NoC router [#2] with MACS router [#1] in terms of area utilization (number of slices) and maximum clock frequency (MHz). Inference from Figure 1.9, the modified

NoC router is shown improved performance because of PE clustering and MDRRA. The improvement in area utilization is small when compared with maximum clock frequency because of IOBs used in each port of the NoC router. The 58% improvement shown in terms of clock frequency compared with MACS router.

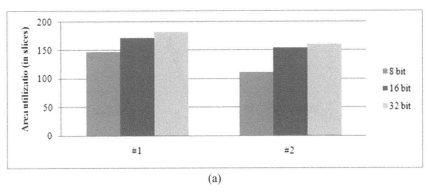

(a)

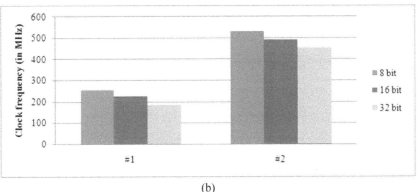

(b)

FIGURE 1.9 Comparison of clock frequency of NoC switch (#2) with MACS switch (#1) for data of 8, 16 and 32 bit.

3×3 mesh-based NoC communication architecture is composed of 9-improved routers and 18 PEs is simulated at Riviera-Pro for windows version. The simulation is conducted for 64 iterations at uniform-random and transpose traffic patterns thereby the average packet latency and bandwidth utilization are observed. Each simulation showed the lower packet latency and higher bandwidth utilization than MACS NoC communication architecture. Figure 1.10 depicts the performance comparison between

3×3 mesh-based NoC and MACS NoC. It is evident that the average data packet latency and bandwidth utilization are improved more than MACS because of advanced architecture of each NoC router with PE clustering and DRRA for arbiter module.

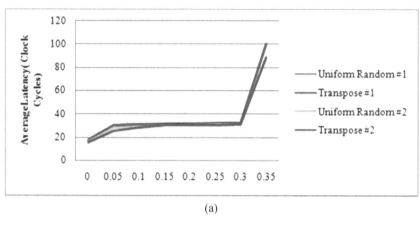

(a)

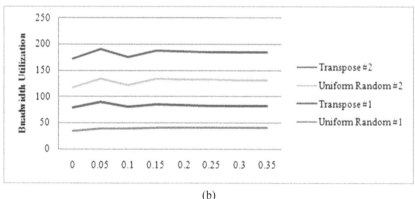

(b)

FIGURE 1.10   Simulation results in terms of average latency and bandwidth utilization of 3×3 mesh-based NoC (#2) MACS switch (#1) at uniform-random and transpose traffic.

Figure 1.11 depicts the evaluation of performance for 4, 8, 16-node parallel OCDMA crossbar switch of NoC router. Figure 1.11(a) shows the increasing area utilization with increasing data packets because of decoder circuit requires a number of parallel blocks to reconstruct original data packets. Figure 1.11(b) represents the speed of parallel OCDMA crossbar in terms of maximum clock frequency and the speed of modified NoC router decreases with increase of number of users because increase in STP and PTS with increase of users. Figure 1.11(c) depicts the total power consumption of NoC router and the power consumption increases linearly with increase of data packet because of increase in the dynamic power consumption with number of STP and PTS modules. Figure 1.11(d) shows the throughput measurement for NoC router.

The throughput (θ) is measured with Eqn. (1):

$$\theta = \frac{N_c N_{bpp} N_{pe}}{t_c} \tag{1}$$

where; N_c is the required number of clock cycles; N_{bpp} is the number of bits at data packet; N_{pe} is the number of received data packets at PE; t_c is the total clock period to complete data packet transfer. The throughput increases with increase in data bits and number of nodes.

Table 1.3 depicts a clear comparison of parallel OCDMA with recent work of CDMA crossbar switch of 8-bit data packet. The comparison is given in terms of area utilization (number of LUT-FF pairs), delay (*ns*) and power consumption (*mW*). Table 1.3 inferred that the parallel OCDMA crossbar switch is showed better results than WB-CDMA [11], SB-CDMA [11] and OCDMA [1] because of parallel execution of encoding and decoding circuit and also adding extra PE to each NoC router. The PE clustering avoids requirement more number of routers to complete the data packet transfer between the source and destination; hence, the data packet delay is reduced. As parallel OCDMA utilizes the orthogonal gold codes, the requirement of multiplexer that selects either orthogonal or non-orthogonal spreading codes at OCDMA eliminated completely; thereby, the performance of proposed work improves. The parallel OCDMA crossbar switch is shown 9.79% reduction of latency (delay) and also shown 20.76% improvement of power consumption when compared with OCDMA.

Figure 1.12 depicts the analysis of data packet latency and throughput for the 3x3 mesh-based NoC communication architecture. For this analysis, the proposed work is simulated at various traffic patterns on Riviera-Pro for windows version. The number of experiments is conducted for 3x3 mesh-based NoC fabric where each NoC router has 2PE and conventional NoC with single PE to investigate the performance metrics. Each experiment of the proposed work is shown lesser data packet latency and higher throughput than the conventional NoC. From Figure 1.12, it is clearly understood that the performance of 3*v*3 mesh-based NoC increases with the increase of number of PEs for advanced router.

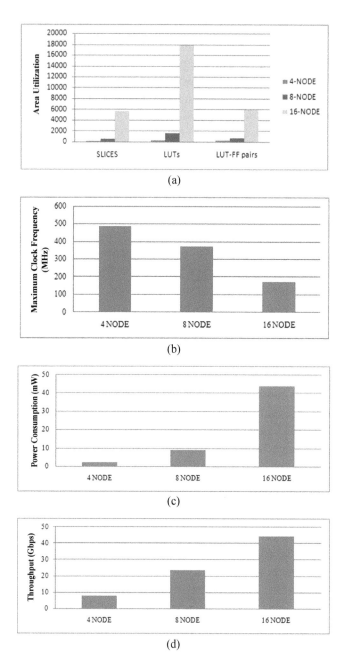

FIGURE 1.11   (a–d) Implementation results in terms of area utilization, maximum clock frequency, power consumption and throughput for NoC with parallel OCDMA crossbar of 4, 8, 16 nodes.

TABLE 1.3 Comparison of Parallel OCDMA with CDMA Crossbar of 8-bit Data Packet Per Switch

CDMA(8-Node)	Area (No. of LUT-FF)	Delay *(ns)*	Power Consumption *(mW)*
WB-CDMA [7]	782	2.82	17.53
SB-CDMA [11]	684	2.71	15.21
OCDMA [1]	692	2.96	11.46
Parallel OCDMA	663	2.673	9.08

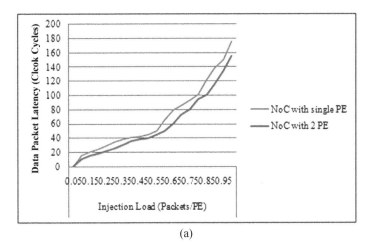

(a)

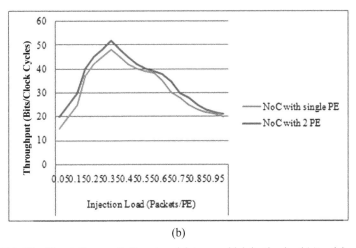

(b)

FIGURE 1.12 Simulation results for network latency with injection load (a) and throughput with injection load (b) in uniform-random traffic pattern.

1.5.1   WHY IS NOC COMMUNICATION ARCHITECTURE NECESSARY FOR IoT APPLICATIONS?

Internet of Things (IoT) is the identification for a wide range of applications and technologies. This advanced technology directs the small-scale system with a large number of processors for applications of short-range communications. The IoT is popularized because of various applications like smart cities, industries, health, home, and transportations. Most of the applications are embedded different hardware technologies thereby maintaining of these hardware technologies for IoT is much complex if the system has less number of processors. The intercommunication of processors is a critical issue when the system has a large number of processors. Hence, new communication architecture is required for multiprocessor System on Chip (MPSoC). NoC is the new paradigms for solving the critical issues when scaling of technology.

1.5.2   FURTHER RESEARCH DIRECTIONS

The data security of NoC communication is one of the important directions for IoT-based applications. The efficient fault-tolerant routing algorithm is also a key constraint for NoC fabric.

1.6   CONCLUSION

The IoT-based applications need powerful communication architecture to compatible performance of PE of system. NoC is the new communication protocol for SoC to resolve critical issues of small-scale system. In this chapter, the architectural parameters of NoC are advanced to improve the performance of SoC. The structure of the arbiter component is improved that MDRRA is proposed to resolve arbitration among ports and also reduce the arbitration delay. A parallel OCDMA is proposed to improve the data transfer speed for crossbar switch of NoC router. By advancing these architectural parameters, the performance 3×3 mesh-based NoC communication architecture was evaluated and compared with recent literature.

KEYWORDS

- **arbiter**
- **crossbar**
- **FSM**
- **network interface**
- **network on chip**
- **router**
- **routing algorithms**
- **topology**

REFERENCES

1. Ahmed, K. E., Rizk, M. R., & Farag, M. M., (2017). Overloaded CDMA crossbar for network-on-chip. *IEEE Transactions on Very Large-Scale Integration Systems, 25*(6), 1842–1855. IEEE.
2. Bashizade, R., & Sarbazi-Azad, H., (2015). P2R2: Parallel pseudo-round-robin arbiter for high performance NoCs. *Integration, the VLSI Journal, 50*, 173–182. Elsevier.
3. Bobda, C., & Ahmadinia, A., (2005). Dynamic interconnection of reconfigurable modules on reconfigurable devices. *IEEE Design & Test of Computers, 22*(5), 443–451. IEEE.
4. Dally, W. J., & Towles, B., (2001). Route packets, not wires: On-chip interconnection networks. In *Proceedings of the 38th Annual Design Automation Conference*, 684–689. ACM.
5. Halak, B., Ma, T., & Wei, X., (2014). A dynamic CDMA network for multicore systems. *Microelectronics Journal, 45*(4), 424–434. Elsevier.
6. Jara-Berrocal, A., & Gordon-Ross, A., (2009). SCORES: A scalable and parametric streams-based communication architecture for modular reconfigurable systems. In: *Proceedings of the Conference on Design, Automation, and Test in Europe* (pp. 268–273).
7. Kim, D., Kim, M., & Sobelman, G. E., (2004). CDMA-based network-on-chip architecture. *Proceedings of the IEEE Asia-Pacific Conference on* in *Circuits and Systems, 1*, 137–140.
8. Kumar, K. A., & Dananjayan, P., (2017). A survey for silicon on-chip communication. *Indian Journal of Science and Technology, 10*(1), 1–10.
9. Kumar, R., & Gordon-Ross, A., (2016). MACS: A highly customizable low-latency communication architecture. *IEEE Transactions on Parallel and Distributed Systems, 27*(1), 237–249. IEEE.

10. Shin, E. S., Mooney, III. V. J., & Riley, G. F., (2002). Round-robin arbiter design and generation. In: *Proceedings of the 15th International Symposium on System Synthesis* (pp. 243–248). ACM.

11. Wang, J., Lu, Z., & Li, Y., (2016). A new CDMA encoding/decoding method for on-chip communication network. *IEEE Transactions on Very Large-Scale Integration Systems, 24*(4), 1607–1611. IEEE.

12. Wang, X., Ahonen, T., & Nurmi, J., (2007). Applying CDMA technique to network-on-chip. *IEEE Transactions on Very Large-Scale Integration (VLSI) Systems, 15*(10), 1091–1100. IEEE.

13. Wiklund, D., & Liu, D., (2002). Design of a system-on-chip switched network and its design support. *IEEE International Conference on Communications, Circuits, and Systems and West Sino Expositions, 2*, 1279–1283.

14. Zheng, S. Q., & Yang, M., (2007). Algorithm-hardware codesign of fast parallel round-robin arbiters. *IEEE Transactions on Parallel and Distributed Systems, 18*(1), 84–95. IEEE.

15. Wu, C. W., Lee, K. J., & Su, A. P., (2018). A hybrid multicast routing approach with enhanced methods for mesh-based networks-on-chip. *IEEE Transactions on Computers, 67*(9), 1231–1245. IEEE.

16. Wang, L., Liu, L., Han, J., Wang, X., Yin, S., & Wei, S., (2019). Achieving flexible global reconfiguration in NoCs using reconfigurable rings. *IEEE Transactions on Parallel and Distributed Systems, 31*(3), 611–622. IEEE.

17. Xu, C., Liu, Y., & Yang, Y., (2019). SRNoC: An ultra-fast configurable FPGA-based NoC simulator using switch-router architecture. *IEEE Transactions on Computer-Aided Design of Integrated Circuits and Systems*, 1–14. IEEE.

18. Kumar, A., & Dananjayan, P., (2019). Parallel overloaded CDMA crossbar for network on-chip. *Facta Universitatis, Series: Electronics and Energetics, 32*(1), 105–118.

19. Kummary, A. K., Dananjayan, P., Viswanath, K., & Karunakar, V., (2019). Combined crosstalk avoidance code with error control code for detection and correction of random and burst errors. In: *Error Detection and Correction* (pp. 1–10). IntechOpen.

20. Ali, H., Tariq, U. U., Liu, L., Panneerselvam, J., & Zhai, X., (2019). Energy optimization of streaming applications in IoT on NoC based heterogeneous MPSoCs using re-timing and DVFS. *IEEE SmartWorld, Ubiquitous Intelligence & Computing, Advanced & Trusted Computing, Scalable Computing & Communications, Cloud & Big Data Computing, Internet of People and Smart City Innovation*, 1297–1304.

CHAPTER 2

Internet of Things (IoT) in Education

T. SENTHIL,[1] J. DEEPIKA,[2] and C. RAJAN[3]

[1]*Department of Electronics and Communication Engineering, Bannari Amman Institute of Technology, Tamil Nadu, India, E-mail: t.senthilece@gmail.com*

[2]*Department of Information Technology, Bannari Amman Institute of Technology, Tamil Nadu, India*

[3]*Department of Information Technology, K. S. Rangasamy College of Technology, Tamil Nadu, India*

2.1 INTRODUCTION

Traditional gaining knowledge of strategies is being changed with agile, collaborative, and technology-primarily based totally training. Education with the most up-to-date school room technology permits present-day gaining knowledge of environments, bridging the virtual divide, and putting together college students for the future. Internet of Things (IoT) in hand with Internet does this wonder. A top-notch technological switch is being achieved with IoT. IoT is withinside the system of transmuting many fields of our ordinary lives. IoT is utilized in all sectors, linking the Internet with gadgets like computers, smartphones, and each different gadget. As Internet has rooted itself into our schools, E-getting to know is becoming a now no longer unusual place exercising throughout the globe. However, the packages of IoT are several in the training quarter, and additionally, its implications are top-notch tremendous [17, 23]. With the upward thrust of Mobile Technology, IoT lets instructors enhance their

Harnessing the Internet of Things (IoT) for Hyper-Connected Smart World.
Indu Bala and Kiran Ahuja (Eds.)

coaching methodology, mode of coaching, efficiency, and get admission to statistics throughout resources. Teachers can use those present-day technologies to convert their conventional elegance room coaching to 'Smart Classroom Teaching.'

2.2   FEW ADVANTAGES OF USING IoT IN THE EDUCATION SECTOR

IoT is a technique for several devices to "communicate" or work together with each other. It renovates precise tasks into fully autonomous tasks.

2.2.1   STAYING CONNECTED

IoT, as a technology transfer, can help teachers and students stay connected to each other. This works out well in the distance mode of education where students and teachers are not expected to give their physical presence [1]. IoT connects people through devices and Internet. An online teaching-learning process can be implemented by any educational institution with IoT.

2.2.2   SECURITY

Despite any domain, security stands as a leading concern. Identifying the student's location is still at the head of teachers mind virtually. Teaches mind always wonder about the top-quality security systems. Yes! This is under the influence of the IoT. Even in schools and colleges, IoT finds its application to keep track of students, for example, by using RFID chip trackers. We can easily trace students' location with the help of RFID chip. Figure 2.1 shows the RFID and Smart Card-based autonomous system for registering students' attendance.

2.2.3   CONTENT BEYOND BOOKS

With IoT, students need not wait for textbooks from schools and colleges. Especially for college students, depending on textbooks is normally not advised. They are yet to explore a lot much about the actual content of the book. Nowadays, preferences are given to different websites to assist student's academics rather than textbooks. This will ease to complete their assignments, overnight study, and even bring your notes together [7]. Another advantage is that students need not worry about the price tag of textbook, rather they would enjoy learning using this emerging technology [8].

FIGURE 2.1   RFID and smart card-based automatic system for schools and colleges.
Source: https://www.youtube.com/watch?v=cq3cFuvTYLs.

2.2.4   BOON TO DIFFERENTLY CHALLENGED

IoT can also help those who have special needs! One of the most recent ways it is used is to translate sign language into adequately written English. That will make it easier for those who rely on it (such as students who are deaf, hard of hearing, or mute) [6]. Figure 2.2 shows the IoT-enabled wheel chair for a differently-abled child.

2.2.5   CLASSROOM MONITORING

IoT in the Education Sector is useful to manage the classroom. This reduces another burden from the hands of a teacher. Instead of thinking about collecting or grading homework, monitoring attendance, and a swing of other things around the classroom that consumed high time is replaced by IoT. IoT supports automatic grading system, attendance monitoring and mark analysis in no matter of time. Thus, teachers can focus on what they are supposed to teach and how can they transfer the concept to students [25].

FIGURE 2.2   IoT helps differently-abled students to enjoy learning.

2.3   UTILIZATION OF IoT IN THE TEACHING SECTOR AND WHY IT IS A GOOD IDEA?

IoT is still in infant stage when we think about education sector. However, this technology is evolving over years in educational institutions. With the help of the Internet, many new forms of cooperative intelligence are being automated [2].

2.3.1   THE SIGNIFICANCE OF RENOVATING THE LEARNING SECTOR

Regardless of important high-tech growth over decades, the implementation of smart classroom technology is believed as a change from the traditional teaching methods. The school syllabus has been intensely modernized by emphasizing much on STEM (science, technology, engineering, and mathematics) courses [12]. We accept that education diminishes poverty, increases financial growth, and rises revenue. But employing current and newer education practices one can reinforce massive growth for a country's educative percentage [12]. Consequently, the digital technological developments have to be integrated with the formal educational system. Or else, it is at the threat of dropping behind [22]. Transformation of the edification segment to a newer perspective can rise the production of own as well as the nation. In short, as the empowerment in the teaching progresses, so too does the intelligence of the pupils [21].

A modern study made by Microsoft and YouGov expressed that 6 out of 10 parents existed happy with the keen laboratory technology-based teaching rather than traditional teaching methods [24]. More intuitively, 86% of parents said that the use of PCs and the added edifying software were extremely useful to their child's edification [20].

2.3.2 CLASSROOMS ARE CHANGING

The use of mobile phones in classrooms are no more a crime nowadays. The usage of electronic devices are accepted among schools as education approaches are targeting to be highly communicating and appropriate to the modern digital world. Technological innovations, such as IoT, artificial intelligence (AI), augmented reality (AR), and virtual reality (VR) are moving distinctive principles of traditional teaching [17, 18]. Figure 2.3 shows the IoT devices that are AR technology embedded for educating school children in classrooms. They are transporting modifications into the normal form. Universities and academies together are accepting an entire innovative world of technological concepts [10].

FIGURE 2.3 IoT and augmented reality (AR) in classroom education.
Source: https://totse.info/smart-education-system.

2.3.3   IMPLEMENTING IoT SYSTEMS IN CLASSROOMS

IoT is a promising technology to reform the classical education organization. Figure 2.4 depicts the smart boards with IoT-enabled feature used for modern education system. Inter-connectivity is connecting previously indifferent offline stuffs and devices. Its programs are numerous, beginning from school rooms or even its extending past the sky. IoT-enabled training answers are used beginning from clever forums to high-school protection programs, thereby offering a smooth bodily surroundings for studying [16]. A technically-modern schoolroom may be provided with the subsequent IoT systems: Communicating whiteboards, automatic attendance monitoring systems, clever ID cards, temperature, and green sensors, clever-green lighting fixtures and predictive renovation for infrastructure, clever heating, ventilation, and air conditioning (HVAC) systems, Wi-Fi door locks and lockdown protocols and protection systems [15].

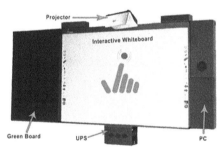

FIGURE 2.4   Smartboards for education.
Source: Mun et al. ICEED-2016. https://www.biz4intellia.com/

2.4 BENEFITS OF IoT IN EDUCATIONAL INSTITUTIONS

Though IoT era has been followed in lots of various industries, packages in faculties are not applied promptly. Meanwhile, the era gives many possibilities for more secure faculties and stepped forward scholar outcomes. The following are the exquisite key blessings of IoT in instructional institutions.

2.4.1 OFFERING STUDENTS SAFETY WITH IoT

The network of IoT is very large interconnecting many nodes such as colored lights, digitized signs, door locks and even sensors [27]. This provides a customized safety for school students. Many schools adopt the technology to create various programs such as weather monitoring, authentication, and intrusion detection and other security risks. In case of emergency, IoT technology enables safety solutions by triggering alert alarms in classrooms and even hostel residence. This affords a safer environment for school students and teachers can ensure their students safe.

2.4.2 IMPROVING STUDENT OUTCOMES WITH IoT

IoT-enabled education is a better way to improve students learning experience says a survey. IoT deployed campus helps students to experience in a much better educational setting. Compared to the traditional education system, IoT campus makes students to involve in learning with interest [26]. Psychologically, IoT attracts students to achieve their learning outcomes with ease.

2.4.3 ENERGY EFFICIENCY AND COST SAVINGS WITH IoT IN SCHOOLS

IoT linked gadgets are programmable and are automated. For example, allowing or disabling Wi-Fi connection withinside the school rooms is achieved with the sensor information linked to programmed software program. Once the scholars input the classroom (the primary pupil), sensors capture this and send a sign to the software program to allow

Wi-Fi and if no pupil is withinside the classroom, the sensor sends the sign to the software program to disable Wi-Fi. IoT connectivity improves constructing performance and decreases electricity waste, ensuing in fee savings.

2.5   IoT APPLICATIONS IN EDUCATION

IoT finds numerous applications in the education sector. Figure 2.5 reveals the most significant features that are to be considered with an IoT-enabled education system.

FIGURE 2.5   Applications of IoT in education.
Source: www.educba.com.

2.5.1   *IoT IN ENERGY MANAGEMENT*

In energy management, IoT plays a major role of resource utilization with the help of sensors installed in devices like cameras, lights, panels, etc.

IoT-based solar panels are used in educational institutions for converting solar energy into electricity. This obviously reduces the power consumption and also maintains a green campus as no carbon dioxide is emitted to the environment. IoT-based campus monitoring system is used as the best IoT solution with the help of a smart grid. Energy efficiency, water consumption and water recycling plant are measured with IoT ecosystem thus launching the strong schooling atmosphere.

2.5.2   SECURITY AND REMOTE ACCESS MANAGEMENT

An alarming aspect of all parents is the security of their children in the educational place. A secured atmosphere for students can be accomplished by choosing IoT ecosystem in schools. Embedded technologies like Near Field Communication (NFC) tags aid to supply students with special "hidden" information. With the help of this, educators can easily monitor students and their access to various parts of the campus such as laboratories and other places within the educational institute [9]. A real-time school room manipulate mechanism may be mounted with NFC that can supply school room registration facts and show the popularity of the schoolroom on LCD panels.

2.5.3   STUDENT HEALTH MONITORING

New technology intentions to blast academic and gaining knowledge of opportunities into the 21st century. Many IoT-based wearable devices are aimed to help student community rather than entertainment alone [3]. With the help of these wearable devices as in Figure 2.6, student's health can be frequently checked and can sense physiological signals over intervals. These devices read the body changes like temperature, heart rate and even signs of depression. Any abnormal values in these signals makes us sure that the particular student is in some rigorous state requiring immediate attention [13]. The system further studies different student's health data such as their medical history, blood pressure, and prescription to detect any signs of depleting health and inform the staff and their parents by raising an alarm/messages through mobile apps.

FIGURE 2.6   IoT embedded wearable devices for health monitoring.
Source: https://iotdesignpro.com/articles.

2.5.4   EDUCATIONAL APPLICATIONS AND INTERACTIVE LEARNING

Many educational mobile apps provide user customization. With the abstracted knowledge from classroom teaching, students can customize their course of action. Students can access their class contents at any time even when on leave. The accessibility is made easy that they can access these contents from their home. Students can create an interactive learning with 3D visualization and animations in their customized text books [4]. Even in classrooms copying the contents from boards or taking notes on a paper are replaced with the advancement in IoT. Instead of writing on text or typing using keypads, students can read out loud, and a voice-based application converts the speech to text and saves them in the digital notebook.

2.6   CHALLENGES OF IoT IN EDUCATION

Despite many advantages of IoT in education, educational institutions need to widely known the traumatic conditions of IoT. Education experts will need to overcome the following problems in advance than setting up IoT in pre-schools, classrooms, and lecture halls.

2.6.1   IMPLEMENTATION COST

Establishing IoT answers for instructional establishments calls for good-sized hardware and software program power. A described IT Infrastructure is essential so that it will set up a custom designed platform collectively with related devices. Hardware, allow bills and maintenance fees are the important factors that intensify the price of an IoT primarily based totally product. Regrettably, each college cannot have the funds for such a highly-priced improvements and layout custom IoT answers.

2.6.2   IN-CLASS ETHICS

Although many new services are in IoT-enabled platform for college kids in dealing with assignments and different stuffs, they may be to be in any such manner of stopping any mode of cheating, piracy, or delivered varieties of educational dishonesty. Before deploying IoT-primarily based totally information sharing systems, a more secure technological platform must be designed for preventing fraud, thereby making sure all shared information is tamper-proof.

2.6.3   LACK OF DATA PROCESSING INFRASTRUCTURE

During the implementation of IoT answers in any academic institution, choice of a dependable computing platform and records equipment are mandatory. The infrastructure that sure establishments enforce for records garage is absolute and risky to empower related IoT-primarily based totally answers.

2.6.4   SECURITY AND PRIVACY CONCERNS

Data accumulation and processing several varieties of virtual facts will place academic establishments at the map for hacking threats [5]. Before deploying an IoT solution, project stakeholders want to collect an eventuality plan for facts gaps, protection attacks, and special threats. Collective attention approximately the importance of facts protection amongst college university college students is a crucial a part of the innovation implementation process [19].

2.7 FEW IoT PROJECTS FOR EDUCATION

In schooling field, a wonderful listing of IoT tasks take their excessive function on the IoT market. Internet of Things tasks for students can streamline the getting to know procedure and enhance coaching procedure as well.

2.7.1 EDMODO

EdModo platform (as in parent 2.7) is designed for mother and father, instructors, and college students, and it is far orientated for colleges and faculties predominantly. It gives announcement, collaboration, and the training capabilities. The community helps instructors to percentage approximately a brand-new material, assignments, checks, and questionnaires. Also, instructors can speak with mother and father and college students in addition to different instructors (Figure 2.7).

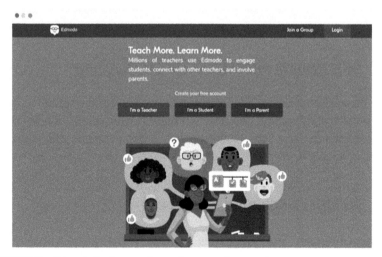

FIGURE 2.7   Edmodo login page for students, teachers, and parents.
Source: www.edmodo.com.

2.7.2 C-PEN

C-Pen brand denotes portable scanners that are to be used by students and teachers. Using C-Pen scanner, one can scan text with a button click and then send this text to the so connected smartphone or other devices. Figure 2.8 shows the device C-Pen that adopts optical character recognition (OCR). C-Pen devices can be improved with a suitable software from a manufacturer.

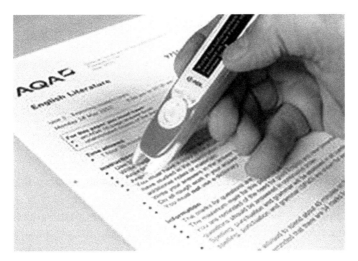

FIGURE 2.8   C-Pen: A digital device that connects to our device.
Source: cpen.com.

2.7.3 NYMI

Nymi wristbands are one of the maximum superior and a hit IoT tasks for today. Some US-primarily based totally faculties already commenced integrating Nymi wristbands into their practice. Nymi wristband (as in determine 2.9) can understand particular coronary heart beating version of every person method that may be used for authentication of a person. The tool blessings instructors in simplifying the roll name as instructors can see proof approximately all college students of their pill. A sign from wristband is going to a pill whilst a pupil enters college or class (Figure 2.9).

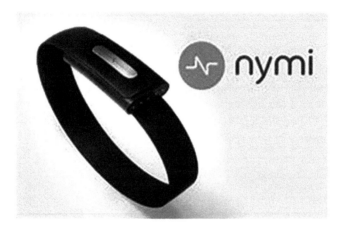

FIGURE 2.9   Nymi Wristbands for or tracking students' authentication.
Source: www.nymi.com.

2.7.4   LOCOROBO

Locorobo (as in decide 2.10) is an IoT application in training that helps programming university college students learn how to format Python, Java, JS, or C apps lots faster. Using an associated tool that learns from its interactions with university college students to optimize the curriculum, human beings of all ages (K-12 and college attendees) can take a look at the basics of robotics, start coding for drones, and format practical wearable applications. There is an associated coding platform and a lesson library that helps STEM teachers get admission to and percent worksheets or lecture notes, similarly to expose the analyzing improvement of each student (Figure 2.10).

2.7.5   MAGICARD

Magicard (as in determine 2.11) permits consolidating all facts needed to provide university attendees with top-notch instructional standards. These clever playing gambling playing cards assist show attendance, offer get right of entry to clinical and transportation facilities, and permit the purchaser to pay for devices withinside the cafeteria at the same time as now not having to maintain coins around. Each tag is, in reality, traceable and may assist decide the student's location. Currently, the subsequent clever card types are available: Payment playing gambling playing cards; Logical get right of entry to playing gambling playing cards; Physical get right of entry to playing gambling playing cards; Time and attendance playing gambling playing cards; Health data gambling playing cards (Figure 2.11).

FIGURE 2.10   LocoRobo.

Source: www.locorobo.co.

FIGURE 2.11   Magicard.

Source: www.magicard.com.

2.7.6   KAJEET

Ensuring steady and comfortable student transportation is a difficulty for optimum community managers. Tools like Kajeet (as in parent 2.12) help beautify the bar for college bus machine efficiency. The platform lets in drivers to show the conduct of passengers onboard; Wi-Fi connectivity lets in college university college students to reveal on homework or get right of get entry to getting to know substances on their manner to school. Kajeet furthermore lets in mother and father to show their child's region in real-time and be confident that they made it to school safely (Figure 2.12).

FIGURE 2.12   Kajeet-mobile hotspot device.
Source: www.kajeet.net.

2.8   EXAMPLES OF IoT TECHNOLOGY IN THE EDUCATION INDUSTRY

Towards integrating IoT in mainstream education, especially schools and universities, the response of all the parties involved is very positive. Many technology companies are providing customized IoT solutions for everything starting from content creation, lesson delivery, and better engagement to classroom and content. A few of them are discussed in subsections.

2.8.1 PROMETHEAN

Promethean is a main organization in interactive shows and lesson shipping software program. The interactive shows allow instructors to layout study room shows in which college students can install their personal thoughts as commonly they want, the use of dry-erase writing. Student engagement has been said to be very excessive whilst those shows are utilized by the instructors. The cloud-primarily based totally dynamic lesson shipping software program permits instructors to adjust instructions in real-time, relying upon pupil response.

2.8.2 BLACKBOARD

Blackboard gives answers and offerings to academic establishments for resolving their maximum crucial and tough problems. From scholar tracking, statistics analytics, and lecture room control to customized getting to know, cellular getting to know and custom-designed getting to know control systems, professionals at blackboard assist establishments leverage era to convert the getting to know ecosystem.

2.8.3 KALTURA

Kaltura video platform affords video advent and control answers to all businesses. Institutions are making getting to know greater interactive, engaging, and on hand the use of Kaltura video answers. It permits instructors and college students to create and manipulate video content material the use of the present-day technologies. These movies may be edited and allotted to any related device.

2.8.4 TYNKER

Coding and programming are believed to force each creativity and technical information in college students simultaneously. Tynker is a coding platform for youngsters elderly 7+. Schools and different academic establishments are the usage of it to ignite the affection of coding of their college students. The self-paced guides are famous amongst mother and father for retaining their youngsters engaged in something innovative and meaningful. Peer interplay presented via way of means of Tynker motivates college students to create remarkable packages themselves.

2.9   THE FUTURE OF IoT IN EDUCATION

IoT gadgets are used to acquire and examine information. Industries starting from production to finance use this clever hardware, which connects to the net and stocks records with different gadgets. The Internet of Things is one of the maximum widespread technological breakthroughs withinside the beyond few years. Having already disrupted some of industries, IoT designers are figuring out a way to convey generation to the lecture room to enhance mastering performance and scholar engagement:

- The advantages of IoT in schooling are endless. Smart generation may want to lessen the want for pencil and paper, assisting school rooms pass inexperienced and reduce down on waste. Plus, there are advantages for college administrators [14]. IoT gadgets can also additionally permit extra statistics-pushed sorts of coaching, in which each college students and educators have get entry to big quantities of information. Students can use these statistics to finish lecture room experiments and study extra approximately subjects like chemistry and physics. At the same time, educators can use statistics on scholar engagement to tell coaching techniques and construct lesson plans. Many school rooms across the United States of America already use IoT gadgets. In clever school rooms, gadgets like ScanMarker and statistics series sensors boom scholar engagement and offer real-time feedback. As this generation will become not unusual place in schools, it will in all likelihood lessen distractions further. The destiny of IoT in schooling will permit new sorts of coaching including fully-far off bodily schooling and tech-nology guides or statistics-pushed coaching techniques. Educators can use this generation to tell methodologies and construct lesson plans. The Internet of Things has a precise rising area in schooling. Here are sure predictions approximately the destiny of IoT.
- By 2025, it is far envisioned that there might be greater than to 21 billion IoT devices.

- More academic sectors will end up "smart."
- Artificial intelligence will hold to end up a larger aspect in coaching learning.
- Routers will hold to end up greater steady and smarter.
- 5G Networks will hold to gas IoT increase in education.

2.10 CONCLUSION

The technological evolution has changed the learning ecosystem in a much interactive way. The traditional method of boring classrooms and lectures are replaced with IoT. Deploying IoT in education is recommended by many psychiatrists as it would transform students mind and keep students involve themselves into the course. IoT devices also provides a safe and secure tracking of students' behavior with RFID tags inside campus. This reduces the teachers overhead on tracking students' movement. Many industries are already into this in developing IoT devices exclusively for education. Modernization of the training region can growth the productiveness of each the character and the nation.

KEYWORDS

- **smart classrooms**
- **safer education**
- **RFID tags**
- **sensors**
- **IoT products**

REFERENCES

1. Brown, J. L., (2017). *How Will the Internet of Things Impact Education?* EdTech Magazine.
2. Burgess, J., Mitchell, P., & Highfield, T., (2017). *Automating the Digital Every Day: An Introduction*. Media International Australia.

3. Byrne, J., O'Sullivan, K., & Sullivan, K., (2017). An IoT and wearable technology hackathon for promoting careers in computer science. *IEEE Transactions on Education, 60*(1), 50–58.

4. Carbonaro, M., King, S., Taylor, E., Satzinger, F., Snart, F., & Drummond, J., (2008). Integration of e-learning technologies in an interprofessional health science course. *Medical Teacher, 30*(1), 25–33.

5. Crawford, K., & Schultz, J., (2014). Big data and due process: Toward a framework to redress predictive privacy harms. *Boston College Law Review, 55*(1), 93–128.

6. Domingo, M. C., (2012). Review: An overview of the Internet of Things for people with disabilities. *J. Netw. Comput. Appl., 35*(2), 584–596.

7. Harris, J., Mishra, P., & Koehler, M., (2009). Teachers' technological pedagogical content knowledge and learning activity types. *Journal of Research on Technology in Education, 41*(4), 393–416.

8. Hollier, S., & Abou-Zahra, S., (2018). Internet of Things (IoT) as assistive technology: Potential applications in tertiary education. *Paper Presented at the Proceedings of the Internet of Accessible Things*. Lyon, France.

9. Hollier, S., (2016). Affordable access. International telecommunications union. *Internet of Things Global Standards Initiative*.

10. Klopfer, E., Sheldon, J., Perry, J., & Chen, V. H, H., (2012). *Ubiquitous games for learning (UbiqGames): Weather Lings, a Worked Example, 28*(5), 465–476.

11. Krotov, V., (2017). The Internet of Things and new business opportunities. *Business Horizons, 60*(6), 831–841.

12. Li, Z., Huang, H., & Misra, S., (2016). Compressed sensing via dictionary learning and approximate message passing for multimedia Internet of Things. *IEEE Internet Things J., 4*, 505–512.

13. Loke, W., (2017). *Healthcare Sector Needs to Evolve* (p. 27). The Business Times.

14. Maksimovic, M., (2017). *Transforming Educational Environment Through Green Internet of Things (g-IoT), 23*, 32–35.

15. Maryam, B. S. H., (2016). The effect of the Internet of Things (IoT) on education business model. In: *12ᵗʰ IEEE International Conference on Signal-Image Technology & Internet-Based Systems.*

16. Meacham, S., Stefanidis, A., Laurence, G., & Phalp, K., (2018). *Internet of Things for Education: Facilitating Personalized Education from a University's Perspective.*

17. Mehmet, A., (2016). *IoT in Education: Integration of Objects with Virtual Academic Communities.* Springer International Publishing Switzerland.

18. Srisakdi, C. P. M., Vo, N. D., & Niek, V. D. L., (2015). Applications of Internet of Things. *Paper Presented at the Twelfth International Conference on eLearning for Knowledge-Based Society*. Bangkok.

19. Ahmad, R. H., & Al-Sakib K. P., (2017). A study on M2M (machine to machine) system and communication: Its security, threats, and intrusion detection system. In: *The Internet of Things: Breakthroughs in Research and Practice*. Information Resources Management Association. IGI Global.

20. Rocha, Á., Correia, A. M., Adeli, H., Reis, L. P., & Mendonça, T. M., (2016). New advances in information systems and technologies. *Advances in Intelligent Systems and Computing, 444.*

21. Rowley, J., (2017). The wisdom hierarchy: Representations of the DIKW hierarchy. *J. Inf. Sci., 33*(2), 163–180.

22. Suja, P. M., & Gondkar, D. R., (2017). Solution integration approach using IoT in education system. *International Journal of Computer Trends and Technology. (IJCTT),* 45–49.

23. Wing, J. M., (2006). Computational thinking. *Communications of the ACM, 49*(3), 3–335.

24. Yuan, J., & Yu, S., (2013). Efficient privacy-preserving biometric identification in cloud computing. In: *INFOCOM, Proceedings IEEE* (pp. 2652–2660).

25. Zain-Ul-Abidin, Farhan, M., Iqbal, M. M., & Naeem, M. R., (2015). Analysis of video lecture's images for the compression formats support for the students in eLearning paradigm. *Sci. Int., 27*, 1171–1176.

26. Zhang, B., Han, J., & Shao, L., (2017). Guest editorial: Feature learning from RGB-D data for multimedia applications. *Multimed. Tools Appl., 76*, 4243–4248.

27. Zhu, Q., Wang, R., Chen, Q., Liu, Y., & Qin, W., (2010). IoT gateway: Bridging wireless sensor networks into the Internet of Things. In: *2010 IEEE/IFIP 8th International Conference on Embedded and Ubiquitous Computing (EUC)* (pp. 347–352).

CHAPTER 3

Innovation in Digital Farming: Relevance in the Present Scenario

ANKUR SINGHAL,[1] TARUN SINGHAL,[1] VINAY BHATIA,[1] DEEPAK DADWAL,[1] and HIMANSHU SHARMA[2]

[1]Chandigarh Engineering College, Landran, Mohali, Punjab, India, E-mail: tarun.sgl@gmail.com (T. Singhal)

[2]J.B. Institute of Engineering and Technology, Hyderabad, Telangana, India

3.1 INTRODUCTION

Since medieval times the human being has been using and exploiting the precious resources given to us by Mother Nature for the fulfillment of our basic needs. Our ancient scriptures like Upanishads and Vedas have clearly mentioned the need to keep the natural resources unadulterated that including atmosphere, sunlight, water, and land, among others. However, with the ever-increasing population has put a lot of burden on them. This is in addition with other factors like urbanization, industrialization has overstrained the asset base and they are getting depleted much more quickly than ever imagined [1–3]. The biggest question in front of the entire world community is how to provide nutritious food and take care of the expanding populace. With the worldwide increase in inhabitants anticipated from 7.6 billion in 2018 to over 9.6 billion out of 2050 there will be a noteworthy increment in the interest for nourishment. Naturally agriculture plays a pivotal role in this regard [3, 4].

Harnessing the Internet of Things (IoT) for Hyper-Connected Smart World.
Indu Bala and Kiran Ahuja (Eds.)

An agrarian network is the composite output of vegetation, soil ripeness, tilling, kinds of yields, earthbound condition, pesticides vegetation, soil ripeness, and so on. Thus, cautious control of all fundamentals required for such a complex interaction is very essential. The main attention is paid to upgrading the crop yield without examining the biological effects of the overexploitation has resulted in ecological imbalance. The efficiency can be increased without disturbing the ecology by increasing the productivity of available resources. The farming and associated sector provides living and occupation to millions of people worldwide. There are in excess of 570 million small scale farmhouses around the globe and farming segment represents 28% of the whole worldwide workforce [3–7]. However, it is expected that in future the accessibility of workforce required for agrarian movement will decrease.

Even according to a recent United Nations finding that global population will touch a figure of 9.7 billion which in comparison is around 34% higher. The main increase in the populace is anticipated in third world countries like Brazil, India, etc., which have the biggest territory on the planet with respect to tillable land for horticulture. To satisfy the growing needs of this populace blast combined with the economic richness of individuals, the worldwide productivity must observe an expansion by 60–70% so that nutritious food may be provided to all individuals. The solution to this overwhelming test requires instantaneous information with a resolute interest towards progress. In this way looking into how traditional agribusiness methods and procedures, while finding new ones, can boost crop productivity, limit ecological effect, and diminish cost is pivotal, presently like never before. Thus, it is necessary to employ novel automation mechanisms to solve unanswered questions regarding the ever-evolving food requirements of billions of individuals.

Thus, it is essentially required to modernize the agriculture system by bringing in much needed reforms by employing technological innovations and bringing them closer to the doorsteps of the farming community. This will help in improving the financial and ecologically economical yield creation. To achieve a global aim of creating a society with zero hunger will necessitate profitable, economical, comprehensive agribusiness systems [2].

So, researchers are keen on how diverse studies of today can help us in the field of farming. This will necessitate a critical change in the present agriculture system. This brings forth the concept of Precision Farming. Modernizations in the digital techniques may be helpful in finding answers to these problems. Fortunately, new advances in the technology have allowed to gather to and influence tremendous volumes of vital information at negligible costs that helps in making farmhouse activities more understanding driven, and possibly progressively profitable and productive. The associated horticulture ecology has started to capitalize on this information technology (IT)-driven expertise. The total scope of the IT-based amenities has grown at a rate of more than 12% in recent years. This brings forth the concept of Precision Farming. The basic definition is utilizing the advanced digital concepts that can effectively manage the agriculture ecosystem and help in analyzing the inconsistencies in the farmlands for optimization of crop production with low investments [6–9]. In simple words, digital farming is a methodology where available resources are used in exact quantities so that agriculture productivity is enhanced in comparison with traditional approaches of farming.

Digital farming is a smart approach that utilizes advances in IT to gather significant information from different places which contributes in the policy formulation. It is expected that diverse conditions will transform digital technological advancements and their utilization in agriculture and allied services that include: accessibility, economic viability and seamless connection, utilization of technology in providing training and formulation of policies. The other contributing factors that further encourage the integration of digital transformation are: speed of internet, cell phones and use of social networking, with effective ICT expertise. It depends on innovations like global positioning systems (GPS), geographic information systems (GIS), harvest screens, geographic information systems (GIS), and sensors with diverse functions for complete observations of farm fields. Greater utilization of advanced agribusiness facilities is imperative in enhancing not only the economic conditions of farmers but also in fulfilling the food requirements of a growing populace. This new situation will require fundamental reexamining by experts, global establishments, business pioneers and the general public with the concept of 'business as usual' can never be an answer [7–15].

3.2 CHALLENGES

The use of digital techniques in agribusiness network includes the hazard that the possible advantages will be inconsistently dispersed between people living in cities and villages. It is anticipated that urban zones have better

'IT-enabled ecological systems' (assets, abilities, and systems) in contrast to countryside regions. This is in addition to higher rate of urbanization with rich communities settling in urban areas, there is possibility that technology divide can worsen the prevailing urban and rural ratio. This can result in rural population unable to understand the process of digitization and get the associated benefits. It is witnessed that the structure and infrastructure of digital system is not developed in villages having enormous native populaces. The expenses related with IT framework present a significant challenge in rustic regions where poverty is on a higher side, particularly in third world countries which are not so developed. In particular, three major challenge issues are: lack of IT system in villages, Rate of IT-enabled literacy and policy framework for precision farming.

3.2.1   LACK OF ICT NETWORK AND SUPPORTING SETUP IN RURAL REGIONS

In this digitization age, IT-enabled services have transformed the world order, including how individuals can gather information, stay connected, and even do trading still there are two worlds in terms of technology usage. Recently all around the world, the footprints of mobile networks are expanding with millions of newer users added into the network. From 2013 onwards, more than 1 billion customers have been connected into the cellular system. A lot of this ongoing development has been witnessed in developing nations. A lot many people around the globe have access to the internet and associated multimedia services with expansion rate higher in developing countries. However, still a large number of individuals remain disconnected, particularly rural folks.

One of the most vital issues is the quality of mobile signals in villages which are still weak in comparison with metro cities. Though with the progression in cellular technology from 2G over to 4G which is already in implementation stage but still subscribers in lease developed countries are unable to have good access to the digital network. It is due to many factors such as conflict areas/war zones in some parts of the globe are also hindering the progression. On the positive side, with the advent of smartphones, subscribers are able to get access to high end web technologies. Along with decreasing expenses of cellular devices and cheaper access plans have made the cellular technology popular worldwide. In spite of the fact that the development of cell phone proprietorship and utilization of IT enabled services has been quicker in under developed nations. However, in comparison, the number of individuals per 100 occupants is higher in developed nations. Still the fundamental hindrance to cell phone possession in developing countries is affordability [16–18].

3.2.2 RATE OF IT-ENABLED LITERACY AMONG RURAL POPULATION

The utilization of advances in IT requires essential education and training particularly to the rural population. Individuals without such capabilities can be deprived of the benefits that come along with the advanced technologies. Generally, in rustic territories, an absence of proper educational and training setup creates hindrance in imparting quality teaching. This prompts decreased interest in educational setup among individuals, particularly the youths with lower participation rates and higher school dropout rates. This impacts in decreased trained workforce again more in rural regions than in urban populace. An absence of fundamental education and expertise develops a huge obstruction in the utilization of innovative techniques [14–19].

Additionally, knowledge of digital technologies is vital for utilizing these advanced innovations. This is in contrast to the situation in developed nations where understudies normally utilize cutting edge digital innovations in their studies and everyday lives, IT skills and aptitudes fall behind in under developed countries. An absence of advanced apparatuses, for example, tablets, and PCs, in schools is distinguished by educators as a significant hindrance to IT instruction. This makes the work of trainers even more difficult resulting in un-utilization of technological benefits.

Rustic joblessness is especially high and excessively influences youth and ladies. The horticultural segment stays a significant source of occupation in villages. Thus, precision farming will altogether revolutionize the whole agriculture sector and its associated work. Progressively, computerized proficiency will be a necessity in agribusiness and its associated employment generation for which appropriate quality training is essential [20–24].

3.2.3 POLICY FRAMEWORK

In numerous nations, governmental policies and agendas plays an influential role in the success of digitalization. This creates a conducive ecosystem for viable merchandise and enabling IT services. There is additionally a pattern by concerned administrators to initialize various e-services for the overall wellbeing of its people and employs digital technologies in crucial sectors such as education, jobs creation, and the health sector. Still, structuring and dealing with an advanced IT-enabled governmental program essentially require a significant level of regulatory mechanism. In some cases, governments are unable to make it successful, particularly in the least developed nations. The underdeveloped nations are frequently the ones with minimal ability to deal with this practice. The success of the governmental programs and policies are segment dependent and for example farming segment a significant source of employment generation sometimes lags. In the next segment, various digital technologies which can be utilized in the agricultural sector are summarized [23–27].

3.3   ROLE OF TECHNOLOGY IN DIGITAL FARMING

The various technologies which can be employed in digital farming are explained in subsections.

3.3.1   GLOBAL SYSTEM FOR MOBILE COMMUNICATION (GSM)

Global system for mobile communication (GSM) has the capability of lossless communication. GSM uses time division multiple access for digitization and compression of data. TDMA uses the same channel at different time slots for transmission of signals. GSM technology offers a better and proficient technique for the high production of crops. Sensors are used to determine the crop parameter. The measured value of parameters like humidity and temperature is compared with some predefined values, and the crop health is notified to the remotely located farmer with the help of GSM technology. The farmer receives crop health at a distance thus reduction in physical effort. The farmer can receive information about crop health from home by a message on phone in at any instant of time and in any weather, condition so can make better utilization of time in production growth. The integration of this technology with traditional methods will result in modernization in agriculture. The most significant obstacle in the production, growth of the crop is the lack of monitoring due to the distance between farmer homes and fields. This obstacle can be overcome by using GSM technology. GSM can be a solution to the agriculture environment monitoring problem. This can alarm farmers timely from upcoming threats. Cheap Internet and a wider range of connectivity can be back born for this technology. Many researchers are working in this field and proposed different cost and energy effective modules based on GSM technology.

3.3.2   GLOBAL POSITIONING SYSTEM (GPS)

Global positioning system (GPS) is a navigation system based on satellite and uses to identify the position of an object at the ground. It was the general belief that digital farming can be realized only by the farmers having large lands and making huge investments in farming, but the simple and low-cost technology like GPS has changed this belief. GPS derived products enhanced the production of the crop. GSM device manufacturer has created so many equipment's to enhance the agribusiness. The main uses of GPS in farming are:

1. **Precision Plowing:** Furrows can be placed in the field with great precision with the help of GPS.
2. **Planting and Fertilization:** Nutrient deficient locations can be traced by the farmers in the field with the help of GPS and the right amount of fertilization can be applied.
3. **Field Mapping:** Exact estimation of field prepared cab be done using GPS.
4. **Machinery Direction:** Machinery has become the necessary part of farming; the location and direction of these machines can be adjusted using GPS for seed placement and distance between seeds.
5. **Yield Monitoring Systems:** Mass flow sensors can be used to monitor the yield in the field by use of GPS.

GPS helps farmers to find wet spots, patches of perennial weeds, drain tile blowouts, and area for the future site. GPS also helps the crop advisors to accurate positioning of pests and insects in the field with the help of rugged data devices.

3.3.3   GEOGRAPHIC INFORMATION SYSTEM (GIS)

GIS is able to capture, store, analyze, and manage data, it is the integration of hardware with software and data. GIS is very helpful in mapping the environmental parameters and crop output. GIS helps to find the best crop to be planted with the help of farming practices in past and soil data analysis. GIS can help in farm asset allocation, field data interpolation, field data reporting, and many more. GIS can be used in a better way with GPS integration.

3.3.4   SENSORS

Digital farming cannot achieve its objective without the use of sensors. Sensors used in digital farming termed as agriculture sensors. Digital farming uses a wide range of sensing technologies that helps farmers in monitoring and optimization of the crop.

Several sensing technologies include in digital farming are:

1. **Location Sensor:** These provide the exact location of an object in terms of longitude and latitude precisely and used to collect location information of a particular area, seeds, and vehicle inside the field.
2. **Optical Sensors:** These types of sensors use the light to collect the information of soil, vehicle, or any other object. These sensors used to find moisture in soil and clay information.
3. **Electrochemical Sensors:** These sensors work on the principle of detecting ions and used to collect chemical data of soil.
4. **Mechanical Sensors:** These sensors work on the principle of contact and probe is used for this purpose.
5. **Dielectric Sensors:** These sensors use the dielectric constant to measure soil moisture.
6. **Airflow Sensor:** These sensors use air permeability to measure different parameters.

The agriculture sensors are used in weather stations for agriculture, agro-industries equipment, Agri drones, mobile operated solar pumps, and in smart fencing to save crops from animals.

3.3.5   INTERNET OF THINGS (IoT)

It is the technology that reliably offers solutions for the advancement of different domains. IoT plays a significant role in digital farming by monitoring of fields with minimum human involvement [6, 7]. Farming can be converted into smart farming by the use of information and communication technologies. In IoT-based smart farming monitoring of crop is done with the help of sensors and irrigation system is automated. IoT-based smart farming is much efficient when compared to conventional methods (Figure 3.1).

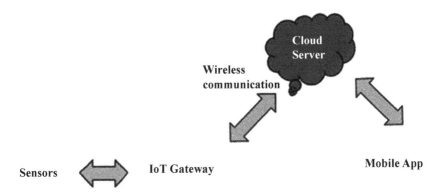

FIGURE 3.1   IoT architecture.

Monitoring of only environmental conditions is not sufficient in the evaluation of production some other factors like the movement of unwanted objects, attacks of wild animals, and theft also affect productivity [9, 10]. IoT architecture is shown in Figure 3.1, the major component in IoT-based farming is the physical structure, data acquisition, data processing, and data analytics. IoT in agriculture provides more advanced solutions when used with other recent technologies like robotics, big data, and many more. Figure 3.2 shows different applications of IoT in digital farming. The IoT applications in farming are:

1. **Agriculture Drones:** The major industry using the drone is digital farming. The drone performs the operation like pesticide spray, assessment of crop health, and monitoring of soil. The drone can also be used in imaging and surveying the field.
2. **Live Stock Monitoring:** The large farm owners can use the wireless IoT application for monitoring their cattle in their large agricultural land. This informs farmers about the sickness of their animals so the preventive measure can be taken timely.

3. **Smart Greenhouse:** Environmental parameters are controlled by the greenhouse manually or automatically. Manual control methods are less effective and increasing labor costs. The design of a smart greenhouse is possible with the help of IoT which gives better control over environmental parameters.

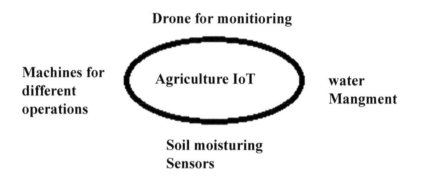

FIGURE 3.2   IoT applications in digital farming.

3.3.6 ROBOTICS

High production with better quality is expected in digital farming at lower expenses. Technology implementation is one of the possible responses to this expectation. Agriculture robots can play an important role in digital farming [11]. These Robots can enhance the speed of work by using multiple techniques. These robots can be controlled with the help of IoT. Robotics in agriculture is a very attractive research domain but still available robots are much slower than humans in the agriculture field. A multi-sensor robotics approach is presented in the field of agriculture robotics [12].

3.3.7 CLOUD COMPUTING AND BIG DATA ANALYTICS

Digital farming is controlling the pressure of increment in food demand and changes in climate. Technical forces like IoT, Cloud Computing, and big data analytics play a crucial role in it. IoT devices play an important role in the collection of data from various sensors plugged in the field and agriculture vehicles. Big Data analytics can compile this data and process it efficiently and can make a better decision with the help of data available in the cloud. Data analytics helps to decide the best time for fertilizer, seeds, and pesticides application.

Few practices where big data analytics makes the difference are:

1. **Feeding a Growing Population:** The demand for the current time is to increase crop production from existing agricultural land. Big data analytics allows the farmer to make smarter decisions in farming and results in increased production.
2. **Using Pesticides Ethically:** Pesticides effects badly to the ecosystem, the use of big data allows farmers to use pesticides in a controlled manner which results in avoid of excess use of chemicals.
3. **Optimizing Farm Equipment:** The use of big data analytics helps the user to know about servicing and other issues of farming equipment's timely and results in the optimized use of farming equipment.
4. **Managing Supply Chain Issue:** Big data analytics can help in tracking and servicing of the delivery vehicle. An efficient delivery system can reduce the wastage of production.

3.4 ROLE OF DIGITAL FARMING IN DEALING WITH COVID-19

COVID-19 has changed the lifestyle of everyone and forces us to change our thinking and conduct of jobs. When Agriculture technology considered, changes we are observing now will become permanent. Data management will play an important role in all types of industries, so agriculture data management becomes important for the agriculture industry. To avoid COVID transmission social distancing plays an important role and we are not touching the objects touched by others. In this scenario, it is important to digitize the records of field parameters in form of agriculture data. Digitization of agriculture data will make it easily accessible to all without any contact with each other. In the last few years, many farmers and agriculture specialists digitized their data but the COVID pandemic will force all to adopt digitization in terms of agriculture data. Digital farming allows us to collect field parameter without any contact with the help of sensors can play an important role in it. In these pandemic situation agronomists are not able to visit the farm by using digital farming all the data can be accessed by agronomists remotely. COVID-19 has created a shortage of manpower in some areas which can be overcome by the use of digital farming. Advancement in robotics in agriculture will outcome in high-speed robots which can reduce the requirement of labor. Agri drones can be used for pesticide spray and fertilization of the field. Digital solutions based on structured data allow farmers to collaborate digitally with their farming partners. Digital applications can be used to access real-time data to increase production. Digital farming allows superior data visibility with the help of technology a particular portion of the field can be accessed remotely. Digital farming also allows farmers to connect digitally with the supply chain so they can get the right cost for their production in this pandemic too.

3.5   FUTURE PROSPECTIVE

Population increasing day by day and so food demand while urbanization decreasing the agricultural land. To compete with the increasing demand for food new technologies required to increase crop production in a smaller area. The concept of digital farming offers précised control overcrop and automation of the farming method and able to fulfill the increased demand. Advancement and availability of current information and communication technologies used in digital farming are its prospects. Development in exiting technology includes better connectivity, new sensors, and superior equipment. Internet connectivity and lack of knowledge is a major issue in the rural area which slows the pace of digital farming. In the future information on the soil will be readily available, so the use of pesticides and fertilizers will reduce and outcomes in the production of the crop with fewer chemicals. Production will increase in the future due to better management of available data on soil and weather conditions. Digital farming can show exponential growth in the upcoming year if supported by favorable policies. R&D efforts and training to young farmers in the digital farming will be the major requirement in the near future.

3.6   CONCLUSION

Agronomy is the science of sowing crops and cattle work. From ancient time Agriculture and associated services played a key role in the ascent of human progress, where the cultivation of good species made quality food accessible to every individual so as to empower them. It also people to live in urban communities. The historical backdrop of agribusiness started a huge number of years prior. Plants were grown in most of the regions independently around the globe. However, with ever-increasing population is has making a lot of strain on agriculture because it is become more difficult to feed quality food to all the individuals. This, along with other challenges like environmental conditions, water depletion, global warming, etc., is has made it mandatory to develop newer technological advancements so as to revolutionize agriculture. Modern digital aids are helping the farming community to increase the crop yield in this century. Innovations in the field of IoT are extensively utilized not only in the devolved nations but also in the developing countries. Still, digital farming brings with their associated problems like lack of quality IT infrastructure and trained manpower, particularly in rural regions. Digital farming is very much need of the hour in present difficult time when the world is facing COVID problem. Everything is lockdown and farmers are unable to physically access their field that frequently. This chapter explains various issues which can be explored with the help of IT technologies.

KEYWORDS

- **global positioning systems (GPS)**
- **geographic information systems (GIS)**
- **agriculture sensors**
- **drones**
- **digital farming**

REFERENCES

1. https://en.wikipedia.org/wiki/Precision_agriculture (accessed on 16 November 2021).
2. FAO, (2017b). *Information and Communication Technology (ICT) in Agriculture: A Report to the G20 Agricultural Deputies*. Rome: FAO.
3. ILOSTAT, (2019). *Employment Database*. Geneva: International Labor Organization.
4. Agfundernews.com. African AgriTech Market Map (accessed on 16 November 2021).

5. Lowder, S. K., Skoet, J., & Raney, T., (2106). The number, size, and distribution of farms, smallholder farms, and family farms worldwide. *World Development, 86,* 16–29.
6. Kamienski, C., & Soininen, J. P., (2019). Smart water management platform: IoT-based precision irrigation for agriculture. *Sensors, 19,* 276.
7. Elijah, O., & Rahman, T., (2018). An overview of Internet of Things (IoT) and data analytics in agriculture: Benefits and challenges. *IEEE Internet Things, 5,* 3758–3773.
8. Zhang, X., & Zhang, J., (2017). Monitoring citrus soil moisture and nutrients using an IoT based system. *Sensors, 17,* 447.
9. Agrawal, H., & Dhall, R., (2019). An improved energy efficient system for IoT enabled precision agriculture. *J. Ambient Intell. Hum. Comput.,* 1–12.
10. Wolfert, S., & Ge, L., (2017). Big data in smart farming: A review. *Agric. Syst., 153,* 69–80.
11. Milella, A., Reina, G., & Nielsen, M., (2019). A multi-sensor robotic platform for ground mapping and estimation beyond the visible spectrum. *Precis. Agricult., 20,* 423–444.
12. Nandhini, A., Hemalatha, R., Radha, S., & Ndumathi, K., (2017). Web-enabled plant disease detection system for agricultural applications using WMSN. *Wireless Personal Communications,* 1–16.
13. Abdel-basset, M., Shawky, L. A., & Eldrandaly, K., (2018). Grid quorum-based spatial coverage for IoT smart agriculture monitoring using enhanced multi-verse optimizer. *Neural Computing & Applications, 4.*
14. Abrahamsen, P., & Hansen, S., (2000). Daisy: An open soil-crop atmosphere system model. *Environmental Modeling & Software, 15,* 313–330.
15. Ahmed, N., De, D., Member, S., & Hussain, I., (2018). Internet of Things (IoT) for smart precision agriculture and farming in rural areas. *IEEE Internet of Things Journal, 5*(6), 4890–4899.
16. Alahmadi, A., Alwajeeh, T., Mohanan, V., & Budiarto, R., (2018). Wireless sensor network with always best connection for internet of farming. *Powering the Internet of Things with 5G Networks,* 176–201.
17. Wolanin, A., & Camps-Valls, G., (2019). Remote Sensing of Environment Estimating crop primary productivity with sentinel-2 and Landsat 8 using machine learning methods trained with radiative transfer simulations. *Remote Sensing of Environment, 225,* 441–457.
18. Zhai, A. F., (2017). Optimization of agricultural production control based on data processing technology of agricultural Internet of Things. *Italian Journal of Pure and Applied Mathematics, 38,* 243–252.
19. Tzounis, A., Katsoulas, N., Bartzanas, T., & Kittas, C., (2017). Internet of Things in agriculture, recent advances, and future challenges. *Biosystems Engineering, 164,* 31–48.
20. Talavera, J. M., Culman, M. A., Aranda, J. M., & Parra, D. T., (2017). Review of IoT applications in agro-industrial and environmental fields. *Computers and Electronics in Agriculture, 142,* 283–297.
21. Verdouw, C., (2016). Internet of Things in agriculture. *CAB Reviews: Perspectives in Agriculture, Veterinary Science, Nutrition, and Natural Resources, 11.*
22. Sjolander, A. J., Thomasson, J. A., Sui, R., & Ge, Y., (2011). Wireless tracking of cotton modules. Part 2: Automatic machine identification and system testing. *Computers and Electronics in Agriculture, 75,* 34–43.

23. Seyyedhasani, H., & Dvorak, J. S., (2018). Dynamic rerouting of a fleet of vehicles in agricultural operations through a dynamic multiple depot vehicle routing problem representation. *Biosystems Engineering, 171,* 63–77.

24. Pham, X., & Stack, M., (2018). How data analytics is transforming agriculture. *Business Horizons, 61*(1), 125e133.

25. Esonen, L. A., & Teye, F. K., (2014). Cropinfra: An internet-based service infrastructure to support crop production in future farms. *Biosystems Engineering, 120,* 92–101.

26. Na, A., & Isaac, W., (2016). Developing a human-centric agricultural model in the IoT environment. *International Conference on Internet of Things and Applications,* 292–297.

27. Miranda, J., Ponce, P., Molina, A., & Wright, P., (2019). Sensing smart and sustainable technologies for agri-food 4.0. *Comput. Industry,* 21–36.

28. Rayes, A., & Salam, S., (2017). *Internet of Things-From Hype to Reality: The Road to Digitization* (pp. 227–233). Springer International Publishing.

29. Wolfert, J., Srensen, C., & Goense, D., (2014). *A Future Internet Collaboration Platform for Safe and Healthy Food from Farm to Fork* (pp. 266–273). SRII. IEEE.

30. Xu, G., Sun, S., Chen, Y., & Wu, Z., (1999). Meeting the challenge of digital earth. *Journal of Remote Sensing,* 85–89.

.

CHAPTER 4

An IoT-Based Smart Jacket for Health Monitoring with Real-Time Feedback

ANURAG SHARMA,[1] ANSHU SHARMA,[2] and MANI RAJ PAUL[2]

[1]GNA University, Phagwara, Punjab, India,
E-mail: er.anurags@gmail.com

[2]CT Institute of Technology and Research, Maqsudan, Jalandhar, Punjab, India

4.1 INTRODUCTION

Nowadays, the Internet of Things (IoT) helps people to live and work smarter. It refers to the physical devices present around the world. These billions of physical devices are connected to daily internet life for sharing and collecting data. Nowadays, home automation has become very easy with the arrival of the different types of computer chips and controllers that make it possible to turn on and off lights directly from mobile. It has become smarter and responsive, merging the world in digital platforms [1]. Any physical device can be transformed into an IoT platform for controlling and communicating with the help of the internet. It is a huge network of connecting physical devices so that people can easily collect and share any kind of data. In the initial time, computer networking started with the economy, sharing, and accessing information for the different computing resources. Soon with the advancement of TCP/IP protocol internet has evolved tremendously [2].

The advancement of the internet consisted of five phases. The initial phase was to connecting the all computers together. The second phase

Harnessing the Internet of Things (IoT) for Hyper-Connected Smart World.
Indu Bala and Kiran Ahuja (Eds.)

was the rise of the World Wide Web which connects a large number of computers. After that mobile phones come into the hand which enabled them to connect all over the world with the internet [3, 4]. In the next phase, the people joined to social networking for sharing and collecting thoughts. Now the present phase is the advent of the Internet of Things that leads to connecting all physical devices with the internet.

IoT-based Smart Jacket has been designed and developed which consists of smart sensors, i.e., ECG sensor, heart rate sensor, body temperature, and surrounding temperature with humidity, oximeter sensors. The data of all the sensors of the patient has been stored on a cloud. Thus, the information of the patient/person wearing this jacket has been sent to the cloud, i.e., the data of the sensors. This enables the doctors, caretakers, and/or patients to observe the data in real-time and suggest any immediate action or medicine to the patient. Moreover, LCD has been placed on this jacket which enables the patient to monitor body temperature and humidity.

4.2   ELECTROCARDIOGRAPHY (ECG)

Electrocardiogram (ECG) is a non-invasive framework which is used as an investigative gadget of the cardiovascular ailments for the human. ECG is the graph of an assortment of the bioelectric potential (Electrical Activity of the heart) as the human heart throbs with respect to time in daily life. It gives critical information about the human heart condition those utilitarian pieces of the heart and the cardiovascular structure of the human. Nowadays, the ECG test has become very simple in labs, hospitals, and other medical shops [1]. The doctor can easily place ECG leads on the human body for the recording of the ECG signal. Different events of the ECG waveforms that may change according to their comparable patient to such a heart rate variability, that the beats are not typical for same and all the while unclear for different sorts of beats in a time. Inferable from this, sometimes the beat classifiers of signals that perform well on the planning data give a bad view appearing on the ECG waveforms of different patients.

ECG is known as electrocardiography that measures the electrical activity of the heart. ECG is a test that tells the functioning of the heart to the patient. It is generally recorded in the form of pulses (heart beat). In ECG graph P, Q, R, S, and T waves are generated from the electrical activity of the human heart as shown in given Figure 4.1 [5]. It measures the electrical pursuit passes across to the heart. A doctor can easily understand the ECG graph for a normal and irregular heartbeat. ECG can easily detect abnormal heart rhythms, chest pain, and heart problems.

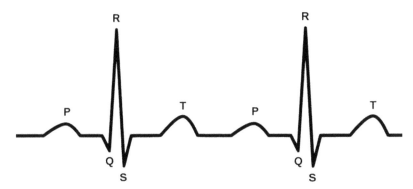

FIGURE 4.1   ECG signal with P, Q, R, S, and T waves.

If the heart rate is too slow, then it is known as bradycardia condition, and if the heart rate is fast, then it is known as tachycardia condition. Both bradycardia and tachycardia conditions can be easily detected with the help of an ECG graph.

As shown in Figure 4.2, the human heart is a solid biological structure of the body in which the blood flows the oxygen in the body in daily life [6]. It gets the unclean blood and deoxygenates that blood through the veins in the human body for filtration. There are four chambers in the human heart. The right ventricle and left ventricle are placed above the portion of the heart while the right ventricle and left ventricle are placed on the lower portion of the human heart [8].

If the upper portion of the heart compresses, then P wave is generated if the lower portion of the heart compresses, then the QRS complex wave is generated, and if the lower portion of the heart comebacks into the normal condition, then T wave is generated [9, 10].

4.3   SENSORS USED IN SMART JACKET

1. **ECG Sensor:** AD8232 Module is used for transmitting ECG signals to IoT cloud. This sensor has three leads as shown in Figure 4.3. Two leads are connected on the chest while one lead is

connected on the left arm of humans. This sensor is controlled by ESP32 controller [11].

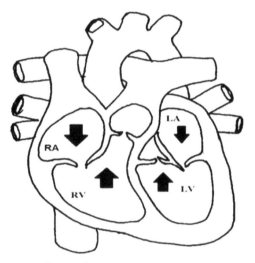

FIGURE 4.2   Structure of human heart.

Source: Modified from: http://clipart-library.com/clipart/piq4je7i9.htm [7].

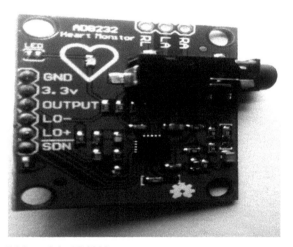

FIGURE 4.3   ECG module AD8232.

2. **Body Temperature Sensor:** DHT11 is used for the measurement of body temperature. This sensor has three pins, as shown in

Figure 4.4. The data of body temperature is also sent to the IoT cloud via Wi-Fi. This smart jacket is suitable to measure temperature in the form of Celsius. This sensor stacks to the human body and starts transmitting data to the cloud [12].

FIGURE 4.4   DHT11 module.

3. **Surrounding Temperature with Humidity:** DHT11 is used for the measurement of body temperature with humidity. Surrounding temperature sometimes is known as ordinary temperature. In this smart jacket body temperature and humidity can also be seen live in LCD if the user has no internet access [12].

4. **Pulse and Oxidation Sensor:** MAX30100 sensor is used for monitoring Pulse and Oxidation levels. A MAX30100 sensor is a small, lightweight device used to monitor the amount of oxygen carried in the body as shown in Figure 4.5. The live data present on the server can be watched live in our android app by the patient and the doctor [13].

5. **Arduino:** It is known as the open-source platform that is used to build and modify electronic projects. Arduino consists of the circuit board and a piece of software as shown in Figure 4.6. The

software of Arduino is used to program the circuit board for many experiments. The software can upload and write the code to the circuit board. In this smart jacket, Arduino UNO is used to display the temperature with humidity values in LCD [14].

FIGURE 4.5   MAX30100 module.

FIGURE 4.6   Arduino UNO.

6. **ESP32:** It is an integrated Wi-Fi and dual-mode Bluetooth controller which is used to control AD8232, DHT11 and MAX30100 sensors in the smart jacket. It is a successor of ESP8266, as shown in Figure 4.7. It can be easily programmed with the help of Arduino software [16].

FIGURE 4.7   ESP 32.

This controller has the ability to connect with the internet and to transfer and receive the values from the cloud. Arduino is known as the open-source platform that is used for building and modifying electronic projects. Arduino consists of the circuit board and a piece of software. In this smart jacket, Arduino UNO is used to display the temperature with humidity values in LCD [16].

7. **LCD:** LCD 16×2 used in a smart jacket for display the temperature with humidity log values. This LCD is connected with Arduino and DHT11 sensor, as shown in Figure 4.8. This LCD has 16 pins and work on 5-volt supply [15].

FIGURE 4.8   LCD 16×2 display [15].

4.4   EXISTING STATE-OF-ART

Most of the research work has been done to design the IoT-based monitoring system to only diagnose single health problems, i.e., ECG. For inventing this innovation, an exhaustive survey of existing literature in all these areas was conducted and a brief review is being presented in Table 4.1.

4.4.1   PROBLEM 1 WITH EXISTING STATE OF ART

In most of the existing systems, the focus has been given to monitor only one problem related to healthcare, i.e., heart problem. There is not any single system that can monitor ECG, body temperature and surrounding temperature, and pulse rate. Real-time data of sensors cannot be monitored by Doctors or caretakers through web-based applications and they cannot revert with immediate action.

- **Solution:** This smart jacket keeps track of a person's ECG, heart rate, body temperature, and surrounding temperature, and pulse rate. This monitoring system stores the data of patients on the cloud and also on an android/web-based application that has been designed for two-way communication between doctors and patients.

4.4.2   PROBLEM 2 WITH EXISTING STATE OF ART

Most of the existing systems use E-mail as a medium to send alert notifications. Sometimes due to lack of the internet, the e-mail is not received properly at the proper time.

- **Solution:** This system sends the data to the server after every second. SMS and e-mail are received by the user when, e.g., the temperature goes above 37.2°C. When the pulse rate becomes normal, SMS, and e-mail will be sent to mobile as an alert message to the users.

The transmission of the ECG signal should be done with the help of IoT cloud, some experts transmit the signal via Bluetooth and Zigbee. But these signals are available within a lower range and do not cover a longer distance. It is possible to send the signal with the help of Wi-Fi, so that

TABLE 4.1 Existing State of Art

SL. No.	Existing State of Art	Drawbacks in Existing State of Art	Sensors Used	Medium	Future Scope
1.	In 2012, Mitra et al. developed a powerful telecardiology application for ECG examination. This technique is the detached weight method and that is suitable for ECG transmission in overall plan of compact (GSM). In this examination (MIT-BIH) arrhythmia database is used for assessment of ECG signal [17].	The data is not used of the real person, only exiting data is used for ECG.	GSM module	GSM	The data of ECG must be used in a real-time of user, then transfer with the help of GSM module.
2.	In 2015, Han Wen et al. developed a weightless and wearable ECG contraption to get an ECG signal. They demonstrated the film catches of 90 seconds to record the ECG indication of the users. This machine is expected to implement four sorts of feeling states [18].	It is basically the offline technique in which the ECG signal is collected from the various subjects. Ninety seconds of a video does not provide any heart arrhythmia. They should have to send ECG signal to the IoT cloud, in which the signal can be analyzed in real-time for emotion states.	ECG	Local	It is a weightless and wearable device, so that the signal should send to the IoT cloud for detecting time and frequency to find the relevant feature.
3.	In 2015, Samuel E et al. developed a wearable device that is suitable for transmission the ECG signal with the help of Bluetooth to the mobile phones [19].	This variable device provides a short-range system because Bluetooth works within a range. The mobile should keep near to the variable device for collecting the output. It does not work when the device is far from wearable system.	ECG	Bluetooth	To overcome this problem of short-range, the ECG signal should be send to the IoT cloud platform so that it can be analyzed anywhere in the world.

TABLE 4.1 *(Continued)*

SL. No.	Existing State of Art	Drawbacks in Existing State of Art	Sensors Used	Medium	Future Scope
4.	In 2015, Brucal et al. developed a portable electrocardiogram capturing device. This device is an Android smartphone that analyze and process the ECG data. This device basically measures heartbeat per minute and RR intervals. The data of ECG is stored into the smartphone through the SD card [20].	In this system the data cannot be seen in real-time, firstly data is recorded with help of Android smartphone and store into the SD card.	ECG	OTG	The data should be sending firstly to the IoT cloud so that it can be seen in a real-time by the doctor and patient. After that download the ECG data and starts the signal processing.
5.	In 2016, Zhe Yang et al. developed IoT-based ECG monitoring system and signal of ECG directly transmit to the IoT cloud. Further, the users and the doctors can access real time data into the web pages [21].	Only the ECG signal is received on the IoT cloud but no other parameter is defined. For the patient, it is very difficult to judge ECG signal without feature extraction and any abnormality present in a signal.	ECG	Wi-Fi	To overcome this problem, the data must be stored on the cloud and then download the data for further analysis to detect feature extraction in ECG signal.
6.	In 2016, M Ryan et al. designed an IoT ECG based transmission web server for monitoring patient ECG. They collect ECG and send raw data directly to the server with the help of Zigbee [22].	This system can handle 20 users at a time and it is available within a shorter range.	ECG	Zigbee	To overcome this problem, the data must be stored on the cloud and then download the data for further analysis to detect feature extraction in ECG signal.
7.	Xinchi Yuer et al. developed a wearable shirt of 12-lead ECG signal, and this wearable shirt is suitable for recording offline ECG signal, they obtained 422 hours of data from five subjects [23]	The ECG signal is received only at the limited range, in other words, it is an offline technique for gathering the ECG data. So, it is available only at the fewer distance	ECG	Wi-Fi	To overcome this problem, the data of ECG must be sent over the IoT cloud. So, anyone in the world can access this ECG data.

TABLE 4.1 *(Continued)*

SL. No.	Existing State of Art	Drawbacks in Existing State of Art	Sensors Used	Medium	Future Scope
8.	Taiyang Wu proposed a real time shirt-based ECG monitoring system and it can send ECG data to users with the help of Bluetooth low energy (BLE) [24].	The data is only sent with the help of Bluetooth low energy so it can be accessed with near devices. It is not used for long-distance it is only available for the under the Bluetooth range.	ECG	BLE	To overcome this problem, the data of ECG must be sent over the IoT cloud. So, anyone in the world can access this ECG data.
9.	Chengy Lou developed a novel IoT-based wearable 12 lead ECG smart vest that is suitable to sending the ECG signal on the cloud. The signal can be accessed by the users in real-time [25].	They only sent the ECG signal to IoT cloud. If the person is does not aware about the ECG signal. It is very difficult to judge for him whether his heart rate is normal or abnormal. No other parameter is described, but only the signal is transmitted with the help of IoT.	ECG	Wi-Fi	To overcome this problem, the data must be stored on the cloud and then download the data for further analysis to detect feature extraction in ECG signal.
10.	Ayaskanta Mishra proposed an ECG monitoring system with the help of AD8232 module using Raspberry pi [26].	Signal analysis should be required for calculating more parameters of ECG signal.	ECG	IoT	Signal processing should be required for calculating ECG compression and feature extraction.

it can cover a longer distance. Experts does not clarify that the signal is received by the receiver is normal or abnormal. So, signal processing is required for processing of signal for analysis ECG parameters.

4.5   IMPLEMENTATION OF IoT MONITORING

All the sensors are used for collecting useful information from the human body. These sensors have different functionality and controlled by the ESP32 controller. After collecting the data from the body, the controller sends the data to the IoT cloud [21]. There are the following parameters that are included in the transmission of data on IoT cloud [27, 28]:

1. **Sensor Module:** The sensors module consists of all sensors, which are used during this project. For the capturing of ECG signal AD8232 model is used. This sensor detects a weak signal from ECG and transmit to the controller. The typical frequency of the ECG signal lies between 0.5 Hz to 100 Hz. For the removal of external noise, the bandpass filter used in the ECG module. For temperature and humidity DHT11 sensor is used and for the pulse rate and oxidation level, a Max30100 sensor is used. All these sensors worked on a 3.3-volt power module.

2. **Controller and Wi-Fi Module:** The ESP 32 controller is used to control all sensors. The programming in ESP32 is done through the Arduino software. Every sensor has different programming for controlling sensors. The Wi-Fi module is inbuilt in ESP32 which is connected to any LAN (Local Area Network) for transmission and receiving of data.

3. **Power Module:** This is generally used for providing the energy to controllers and sensors.

4.5.1   MQTT PROTOCOL

MQTT consists of the four concepts:

- Publish/subscribe;
- Messages;
- Topic; and
- Broker.

The first type of MQTT protocol is to publish and subscribe to the device [21]. In the publishing and subscribe system firstly a device can publish a message to the server it can either be subscribed to a particular topic to receive a message [29]. In a second concept, message contains information that you want to exchange between your device and the server. It can either be a command or data. The third main concept is related to the topics [30]. As shown in given Figure 4.9, complete setup is followed for this protocol.

The topics are very important concept used in MQTT protocol. This protocol specifies interest for incoming messages and defines where we want to publish the data or message. These topics are always defined by strings which is separated by the slash and topics are case sensitive [31, 32].

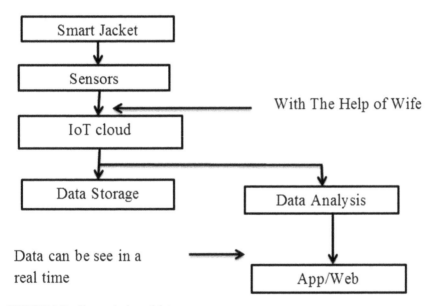

FIGURE 4.9   Transmission of data.

The MQTT broker receives the message as well as it also filters a message. It can also publish the message to the subscribing client. So many brokers can be used for home automation, but the mosquito broker can be easily installed with the help of Raspberry Pi but alternatively, the MQTT broker is used [21]. The data of all sensors are stored on the IoT cloud [33, 34].

4.6   STEPS FOR DEVELOPMENT OF SMART JACKET

- Development of an IoT cloud-based ECG Monitoring system with body, ambient temperature, and oxidation level;
- Feature extraction of the ECG for abnormality detection;
- Development of feedback mechanism to revert an alert message to mobile phones;
- Development of an App/Website in which data can be displayed in real-time.

4.6.1   *DEVELOPMENT OF AN IoT CLOUD-BASED ECG MONITORING SYSTEM WITH BODY, AMBIENT TEMPERATURE, AND OXIDATION LEVEL*

4.6.1.1   *ARRANGEMENT OF THE SENSOR ON THE BODY*

As shown in Figure 4.10, ECG electrodes are placed on the human body. Two electrodes are placed on the chest while the other electrode is placed on the right-hand side of the body. These three electrodes have three leads and these leads are the pair of positive and negative electrodes. The leads can be classified into two types bipolar leads and unipolar leads. Bipolar leads record the potential difference between the positive and negative poles and Unipolar leads records the electrical potential at a particular point. The temperature sensor is placed below the two electrodes this temperature sensor easily records the body temperature and ambient temperature with humidity. On the left-hand side pulse rate and oxidation level sensor is placed.

4.6.1.2   *TRANSMISSION AND RECEIVING OF DATA*

As shown in given Figure 4.11. The smart jacket consists of a sensor for transmitting data to the IoT cloud. The sensors send the log values with the help of Wi-Fi (controlled by ESP32). ESP32 controller must have to connect with the internet for transmitting the data to the cloud.

The data of all sensors has been sent to IoT cloud with the help of the MQTT protocol. As shown in Figure 4.12, the different types of data are obtained on the cloud. The body temperature received on the cloud for user is 33.00 (normally known as 33°C) while the ambient temperature has been recorded is 21.00 (normally known as 21°C).

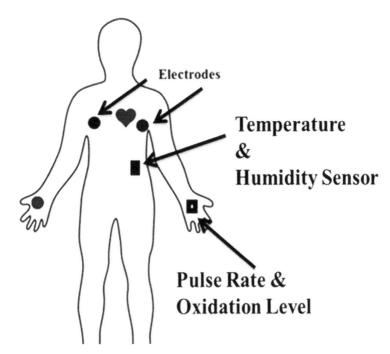

FIGURE 4.10   Arrangement of sensors on body.

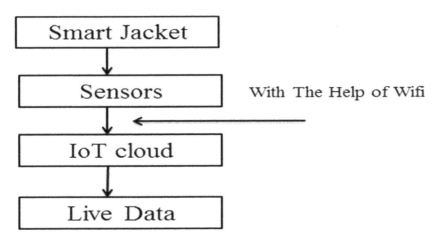

FIGURE 4.11   Transmission of data.

As shown in Figure 4.13, humidity is recorded 84.00 and the value of oximeter is recorded 93.00. As shown in given Figure 4.14, the ECG signal is recorded from cloud and the pulse rate is 87.00 obtained.

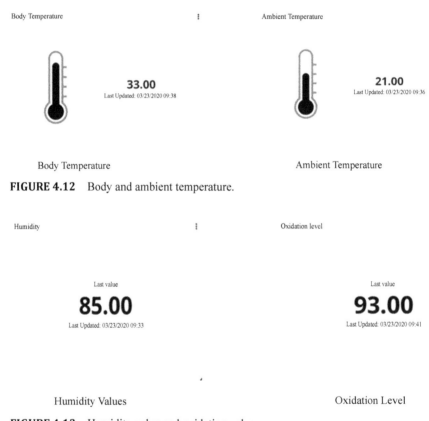

FIGURE 4.12   Body and ambient temperature.

FIGURE 4.13   Humidity value and oxidation value.

4.6.2   *FEATURE EXTRACTION OF THE ECG FOR ABNORMALITY DETECTION*

4.6.2.1   *THE DETECTION OF R PEAK*

In the processing of ECG signal, the recognition of the main QRS complex is a very important obligation for the examination of ECG signal. In QRS complex only the R peak is very significant for the recognition

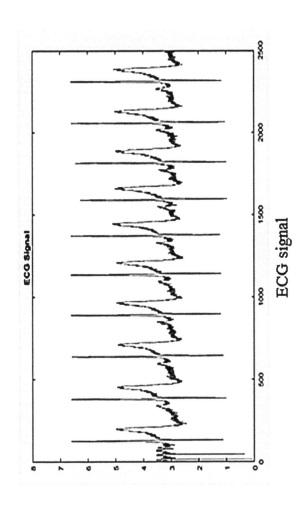

FIGURE 4.14 Pulse rate and ECG signal.

and examination of the ECG signal [35, 36]. As shown in the given figure R-wave from ECG signal is detected with the tool. Sometimes, morphological difference in ECG causes an increase in complexity of the main segment of QRS detection. In ECG there are so many modes and approaches are modified to upgrade the accuracy of the main QRS complex and also the detection of QRS complex [37, 38]. In some cases, a very exact recognition of the main QRS complex is very difficult because ECG signal has different varieties of noise present in a signal like electrode motion and power line interference, etc. Khazaee et al. proposed the power spectral density feature of each heartbeat of the ECG signal that will be the three-timing in intervals picture classifying the cardiac abnormalities that are present in the MIT-BIH database of the ECG signal. Dutta et al. all proposed that the cross correlation-based feature of the ECG signal (Figure 4.15) [39, 40].

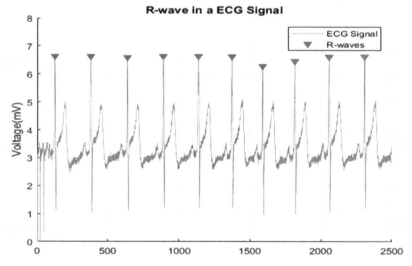

FIGURE 4.15   R peak detection.

As shown in Figure 4.16, a number of speaks has been detected with the help of MATLAB tool for finding heart rate and process the signal for further analysis to detect Heart Rate Variability.

The tool has been developed in MATLAB for finding the heart rate. Just load the ECG signal in tool and this tool automatically shows the heart rate. It also checks the normal and abnormal signal. If the heart will lie between 60–120 beats in one minute, then it is the sign of normal heart beat [41–43]. But if the heart will lie above 120 beats in one minute, then it is a sign of an abnormal heart beat. Figure 4.17 has shown the ECG signal of a 2-year-old child. R peak for 2-year-old child is close with one other. Figure 4.18 shows the ECG signal of 67-year-old person and R peak for 67-year-old person is separate from certain distance.

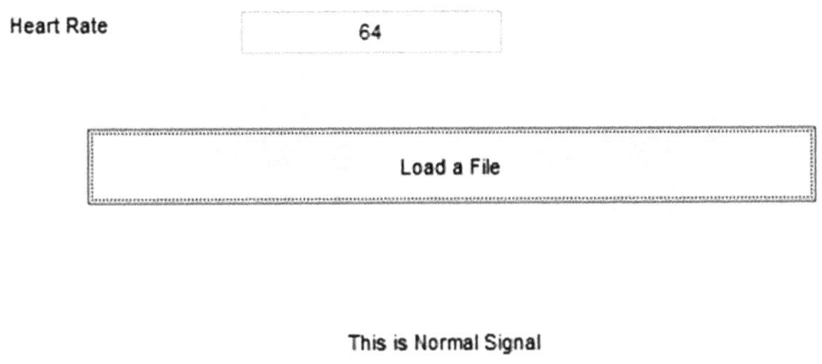

Heart Rate   64

Load a File

This is Normal Signal

FIGURE 4.16   Heart rate tool.

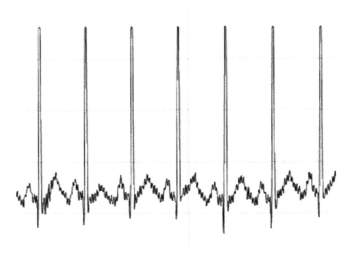

FIGURE 4.17   ECG of 2-year-old child.

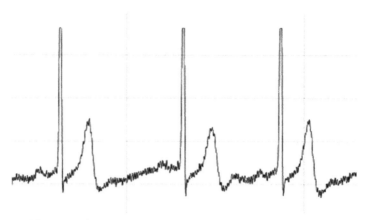

FIGURE 4.18   ECG of 67-year-old person.

4.6.2.2   ECG COMPRESSION

ECG signal compression is one of the most important methods for processing and analysis of ECG signal [44]. Compression helps to reduce the unwanted noise present in a signal and a reduced the size of the signal.

Different parameters are used in signal compression to check the quality of ECG signal [45, 46]. Some of these parameters are namely: compression ratio (CR), and percentage root mean squared difference (PRD). Data compression is classified into three categories such as transformation domain methods, direct data compression methods and parameter extraction methods [47]. As shown in Figure 4.19, compression of ECG signal is obtained.

4.6.3   DEVELOPMENT OF A FEEDBACK MECHANISM TO REVERT AN ALERT MESSAGE TO MOBILE PHONE

4.6.3.1   SETTING UP ALERT FEEDBACK MECHANISM

With the help of IoT cloud, threshold limit has been set for a few sensors on the IoT cloud as shown in Figure 4.20 methodology is followed for this mechanism. Threshold limit for the temperature sensor has been set on IoT cloud at 37°C. Because 37.2°C is a condition, when human feels a fever.

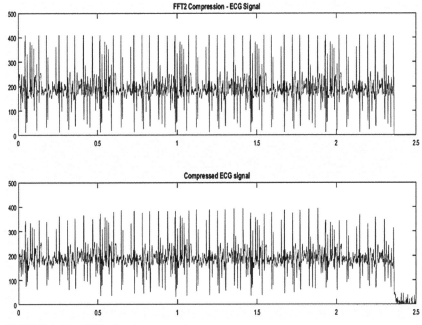

FIGURE 4.19 ECG signal compression.

This smart jacket is also capable to detect the fever and sending the alert message to the doctor, when the patient has a suffering from fever. The threshold limit has been also placed for pulse rate and oxidation level sensors.

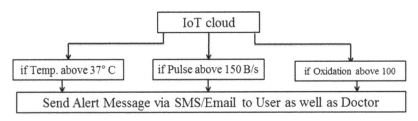

FIGURE 4.20 Alert feedback mechanism.

When the body temperature goes above 37°C that means the user is suffering from fever. As shown in Figure 4.21, SMS is received by a doctor in his mobile. All the log values of all sensors have been seen directly in Android app.

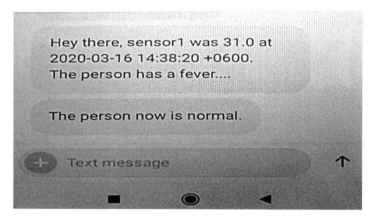

FIGURE 4.21   Alert feed sent by smart Jacket to doctor via message.

4.6.4   DEVELOPMENT OF AN APP/WEBSITE IN WHICH DATA CAN BE DISPLAY IN A REAL-TIME

Android app has very simple GUI design as shown in Figure 4.22 which is suitable for everyone for watching and understanding the log values. With the help of this Android app user and doctor can communicate with each other. A doctor can respond the patient query with the help of this app.

FIGURE 4.22   Android app.

4.7 WORKING OF SMART JACKET

This smart jacket consists of sensors like ECG sensor (ad8232 Module), temperature sensor (DHT11) and pulse and oxidation level sensor (Max30100). These sensors are controlled by ESP32 controller. ESP32 is a controller that is the ability to connect with the internet for transmitting and receiving a real-time data. All sensors are connected with ESP32. These sensors collect the response from the human body and transmitting the log values (data values) to the ESP32. Figure 4.23 describes the jacket overview.

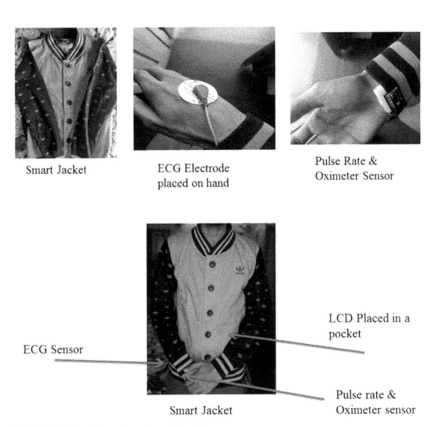

Smart Jacket

ECG Electrode placed on hand

Pulse Rate & Oximeter Sensor

LCD Placed in a pocket

ECG Sensor

Smart Jacket

Pulse rate & Oximeter sensor

FIGURE 4.23 Overview of smart jackets.

Now, ESP32 is connected with the internet and start transmitting the data to the IoT cloud. ESP32 needs a programing code for sharing and controlling the data from sensors as well as from the internet. It is generally programmed in Arduino software which is open-source programming software. ESP32 securely sends the data with the help of the MQTT (Message Queuing Telemetry Transport) protocol to the server. Every server has a unique identification key for exchanging information between two or more physical devices.

4.8   CONCLUSION

The wearable health monitoring device is an important part of our daily life. Nowadays, it is necessary to create a platform for smart healthcare system that is a cost-effective, user-friendly, and self-monitoring system. It provides the people with efficient and effective solutions to live in their homes instead of going to hospitals. This smart jacket fully provides the smart healthcare system. Firstly, ECG signal is recorded from the patient and then parameters are calculated. In a case of emergency, immediate alert message will help to save many lives in the world. A complete health care monitoring system has been developed in which sensors are embedded into the smart jacket. In this smart jacket ECG signal body, ambient temperature with humidity, pulse rate, and oxidation level. This jacket also detects the fever and sends the alert message to the doctor. An Android app has been developed in which real-time data can be seen. The doctor can also communicate with the user through this app. Similarly, the user can communicate and reply to the message to the doctor within this app. This smart jacket is useful for patients for saving their health from any disorder.

KEYWORDS

- electrocardiography
- ESP32
- healthcare
- Internet of Things
- smart jacket

REFERENCES

1. Yang, Z., Zhou, Q., Lei, L., Zheng, K., & Xiang, W., (2016). An IoT-cloud based wearable ECG monitoring system for smart healthcare. *Journal of Medical Systems, 40*(12), 1–11.
2. Guillemin, P., & Friess, (2009). *The Industrial Internet of Things Volume G1: Reference Architecture.* The Cluster of European Research Projects, Technical Report.
3. Atzori, L., Iera, A., & Morabito, (2010). The Internet of Things: A survey. *Computer Networks 54*(15), 2787–2804.
4. Atzori, L., Iera, A., & Morabito, G., (2017). Understanding the Internet of Things: Definition, potentials, and societal role of a fast-evolving paradigm. *Ad. Hoc. Networks, 56*, 122–140.
5. Cromwell, L., Weibell, F. J., & Pfeiffer, E. A., (2014). *Biomedical Instrumentation and Measurements* (pp. 105–125). PHI Learning Private Limited, New Delhi.
6. Acharya, U. R., Bhat, P. S., Iyengar, S. S., Roo, A., & Dua, S., (2003). Classification of heart rate data using artificial neural network and fuzzy equivalence relation. *Pattern Recognition, 36*(1), 61–68.
7. Modified from http://clipart-library.com/clipart/piq4je7i9.htm (License-Personal Use) (accessed on 16 November 2021).
8. Haykin, S., (2002). *Neural Networks* (pp. 14–18). New Delhi: Pearson Education Asia.
9. Rangayyan, R. M., (2001). *Biomedical Signal Analysis: A Case-study Approach* (pp. 18–28). Wiley-Interscience. New York.
10. Ince, T., Kiranyaz, S., & Gabbouj, M., (2009). A Generic and robust system for automated patient-specific classification of ECG signals. *IEEE Trans. Biomed. Eng., 56*(5), 1415–1426.
11. Prasad, A. S., & Kavanashree, N., (2019). ECG monitoring system using AD8232 sensor. *International Conference on Communication and Electronics Systems (ICCES)* (pp. 976–980). Coimbatore, India. doi: 10.1109/ICCES45898.2019.9002540.
12. Margret, S. F., Suryaganesh, P., Abishek, M., & Benny, U., (2019). IoT based smart window using sensor Dht11. In: *5th International Conference on Advanced Computing & Communication Systems (ICACCS)* (pp. 782–784). Coimbatore, India. doi: 10.1109/ICACCS.2019.8728426.
13. Saçan, K. B., & Ertaş, G., (2017). Performance assessment of MAX30100 SpO$_2$/heartrate sensor. *Medical Technologies National Congress (TIPTEKNO)* (pp. 1–4). Trabzon. doi: 10.1109/TIPTEKNO.2017.8238126.
14. Nayyar, A., & Puri, V., (2016). A review of Arduino boards, Lilypad's & Arduino shields. In: *3rd International Conference on Computing for Sustainable Global Development (INDIACom)* (pp. 1485–1492). New Delhi.
15. Metering, A. S., Visalatchi, S., & Sandeep, K. K., (2017). Smart energy metering and power theft control using Arduino & GSM. In: *2nd International Conference for Convergence in Technology (I2CT)* (pp. 858–961). Mumbai. doi: 10.1109/I2CT.2017.8226251.
16. Barybin, O., Zaitseva, E., & Brazhnyi, V., (2019). Testing the security ESP32 Internet of Things devices. In: *2019 IEEE International Scientific-Practical Conference Problems of Info Communications, Science, and Technology (PIC S&T)* (pp. 143–146). Kyiv, Ukraine. doi: 10.1109/PICST47496.2019.9061269.

17. Pal, S., & Madhuchhanda, M., (2015). Electrocardiogram data compression using adaptive bit encoding of the discrete Fourier transforms coefficients. *IET Sci. Meas. Technol., 9*(7), 866–874.

18. Guo, H., Huang, Y., Chien, J., & Shieh, J., (2014). Short-term analysis of heart rate variability for emotion recognition via a wearable ECG device. *International Conference on Intelligent Informatics and Biomedical Sciences (ICIIBMS)*, 262–265.

19. De Lucena, S., Sampaio, D., Mall, B., Meyer, M., Burkart, M., & Keller, F., (2014). ECG monitoring using android mobile phone and Bluetooth. *IEEE International Instrumentation and Measurement Technology Conference (I2MTC) Proceedings.* Available: 10.1109/i2mtc.2014.7151584.

20. Brucal, S., Clamor, G., Pasiliao, L., Soriano, J., & Varilla, L., (2016). Portable electrocardiogram device using android smartphone. In: *38th Annual International Conference of the IEEE Engineering in Medicine and Biology Society (EMBC)* (pp. 509–512).

21. Shaown, T., Hasan, I., Mim, M. M. R., & Hossain, M. S., (2019). IoT-based portable ECG monitoring system for smart healthcare. In: *2019 1st International Conference on Advances in Science, Engineering, and Robotics Technology (ICASERT)* (pp. 1–5). Dhaka, Bangladesh.

22. Nurdin, M., Hadiyoso, S., & Rizal, A., (2016). A low-cost Internet of Things (IoT) system for multi-patient ECG's monitoring. *International Conference on Control, Electronics, Renewable Energy and Communications (ICCEREC)*, 7–11.

23. Yu, X., et al., (2017). A wearable 12-lead ECG T-shirt with textile electrodes for unobtrusive long-term monitoring-Evaluation of an ongoing clinical trial. *EMBEC & NBC,* 703–706.

24. Wu, J. R., & Yuce, M., (2018). A wearable, low-power, real-time ECG monitor for smart T-shirt and IoT healthcare applications. *Internet of Things*, 165–173.

25. Liu, C., Zhang, X., Zhao, L., Liu, F., Chen, X., Yao, Y., & Li, J., (2018). Signal quality assessment and lightweight QRS detection for wearable ECG smart vest system. *IEEE Internet of Things Journal, 6*(2), 1363–1374.

26. Mishra, A., Kumari, A., Sajit, P., & Pandey, P., (2018). Remote web-based ECG monitoring using MQTT protocol for IoT in healthcare. *International Journal of Advanced Engineering and Research Development, 5,* 04.

27. Gayathri, M. S., Ravishankar, A. N., Kumaravel, S., & Ashok, S., (2018). Battery condition prognostic system using IoT in smart microgrids. In: *3rd International Conference on Internet of Things: Smart Innovation and Usages (IoT-SIU)* (pp. 1–6). Bhimtal.

28. Acer, U. G., Boran, A., Forlivesi, C., Liekens, W., Pérez-Cruz, F., & Kawsar, F., (2015). Sensing Wi-Fi network for personal IoT analytics. In: *5th International Conference on the Internet of Things (IoT)* (pp. 104–111). Seoul. doi: 10.1109/IOT.2014.7356554.

29. Yokotani, T., & Sasaki, Y., (2016). Comparison with HTTP and MQTT on required network resources for IoT. *International Conference on Control, Electronics, Renewable Energy and Communications (ICCEREC)* (pp. 1–6). Bandung. doi: 10.1109/ICCEREC.2016.7814989.

30. Sarierao, B. S., & Prakasarao, A., (2018). Smart healthcare monitoring system using MQTT protocol. In: *3rd International Conference for Convergence in Technology (I2CT)* (pp. 1–5). Pune. doi: 10.1109/I2CT.2018.8529764.

31. Wukkadada, B., Wankhede, K., Nambiar, R., & Nair, A., (2018). Comparison with HTTP and MQTT in Internet of Things (IoT). *International Conference on Inventive Research in Computing Applications (ICIRCA)* (pp. 249–253). Coimbatore. doi: 10.1109/ICIRCA.2018.8597401.

32. Almazroi, A. A., (2019). Security mechanism in the Internet of Things by interacting HTTP and MQTT protocols. *IEEE 11th International Conference on Communication Software and Networks (ICCSN)* (pp. 181–186). Chongqing, China. doi: 10.1109/ICCSN.2019.8905262.

33. Kanade, V. A., (2017). "Organic optical data storage" for securely safeguarding IoT secrets. *International Conference on Big Data, IoT, and Data Science (BID)* (pp. 148–153). Pune. doi: 10.1109/BID.2017.8336589.

34. Pramukantoro, E. S., Bakhtiar, F. A., & Bhawiyuga, A., (2019). A semantic RESTful API for heterogeneous IoT data storage. *IEEE 1st Global Conference on Life Sciences and Technologies (LifeTech)* (pp. 263–264). Osaka, Japan. doi: 10.1109/LifeTech.2019.8884026.

35. Sasweta, P., Dash, M., & Sabut, S. K., (2016). DWT-based feature extraction and classification for motor imaginary EEG signals. *International Conference on Systems in Medicine and Biology (ICSMB)*. doi: 10.1109/ICSMB.2016.7915118.

36. Sherathia, P. D., & Patel, V. P., (2017). Sensitivity and positive prediction accuracy analysis for r peak detection in ECG feature extraction. In: *2nd International Conference for Convergence in Technology (I2CT)* (pp. 680–685). Mumbai. doi: 10.1109/I2CT.2017.8226216.

37. Juie, D. P., & Rajveer, S., (2014). Feature extraction of ECG signal. *International Conference on Communication and Signal Processing*. doi: 10.1109/ICCSP.2014.6950168.

38. Aqil, M., Jbari, A., & Bourouhou, A., (2016). Adaptive ECG wavelet analysis for R-peaks detection. *International Conference on Electrical and Information Technologies (ICEIT)* (pp. 164–167). Tangiers. doi: 10.1109/EITech.2016.7519582.

39. khazaee, A., & Ebrahimzadeh, A., (2010). Classification of electrocardiogram signal with support vector machines and genetic algorithms using power spectral features. *Biomedical Signal and Control, 5,* 252–263.

40. Dutta, S., Chatterjee, A., & Munchi, S., (2010). Correlation technique and least square support vector machine combined for frequency domain-based ECG beat classification. *Medical Engineering & Physics, 32,* 1161–1169.

41. De Chazal, P., Duyer, M. O., & Reilly, R. B., (2004). Automatic classification of heartbeat using ECG morphology and heartbeat interval features. *IEEE Trans. Biomed. Eng. 51,* 1196–1206.

42. Dave, T., & Pandya, U., (2018). R-peak extraction for wireless ECG monitoring system. *International Conference on Advances in Computing, Communications, and Informatics (ICACCI) (*pp. 995–999*). Bangalore.* doi: 10.1109/ICACCI.2018.8554750.

43. Malgina, O., Milenkovic, J., Plesnik, E., Zajc, M., & Tasic, J. F., (2011). ECG signal feature extraction and classification based on R peaks detection in the phase space. *IEEE GCC Conference and Exhibition (GCC)* (pp. 381–384). Dubai. doi: 10.1109/IEEEGCC.2011.5752544.

44. Pandhe, D. C., & Patil, H. T., (2014). ECG data compression for a portable ECG recorder and transmitter. *International Conference on Advances in Communication*

and Computing Technologies (ICACACT 2014) (pp. 1–5). Mumbai. doi: 10.1109/ EIC.2014.7230714.

45. Aneesh, K. N., Darshan, S. S., Abhishek, M. H., & Shreekanth, T., (2019). Two-dimensional ECG signal compression based on region of interest using PCA. *International Conference on Communication and Electronics Systems (ICCES)* (pp. 248–252). Coimbatore, India. doi: 10.1109/ICCES45898.2019.9002270.

46. Shinde, A. A., & Kanjalkar, P., (2011). The comparison of different transform-based methods for ECG data compression. *International Conference on Signal Processing, Communication, Computing, and Networking Technologies* (pp. 332–335). Thuckafay. doi: 10.1109/ICSCCN.2011.6024570.

47. Pallavi, M., & Chandrashekar, H. M., (2016). Study and analysis of ECG compression algorithms. *International Conference on Communication and Signal Processing (ICCSP)* (pp. 2028–2032). Melmaruvathur. doi: 10.1109/ICCSP.2016.7754531.

CHAPTER 5

Cognitive Internet of Things, Its Applications, and Its Challenges: A Survey

ROHIT ANAND,[1] NIDHI SINDHWANI,[2] and SAPNA JUNEJA[3]

[1]G.B. Pant Engineering College, New Delhi, India,
E-mail: roh_anand@rediffmail.com

[2]Amity University, Noida, Uttar Pradesh, India

[3]B.M. Institute of Engineering and Technology, Sonepat, Haryana, India

5.1 INTRODUCTION

Internet-of-things (IoT) is a system of various things that can be devices, machines, objects, or people. All of them have the capability to transfer information over a network. Cognitive Internet of Things (CIoT) refers to the use of cognitive computing techniques in the various actions performed by the various self-configured machines connected to each other. In CIoT, the various IoT devices perform five types of cognitive tasks: perception, data analytics, knowledge recognition, decision-making, and service prearrangement. The various things in CIoT can generate and transfer the data by examining the surroundings to make new services [1, 2].

Cognitive IoT is quite useful as it characterizes a smart behavior that may be pervaded into the various mechanisms and devices through cognitive computing and IoT [3]. CIoT increases the intelligence and efficiency of the entire system. It is aimed towards optimizing the performance and behavior of the whole network [4]. Further, a smart CIoT system may be

Harnessing the Internet of Things (IoT) for Hyper-Connected Smart World.
Indu Bala and Kiran Ahuja (Eds.)
© 2023 Apple Academic Press, Inc. Co-published with CRC Press (Taylor & Francis)

created that may be able to generate its own decisions [5]. CIoT system has the potential to explore the available storage space in cloud servers and transfer the data with spectrum sensing.

The cognitive IoT network keeps on searching for the vacant spectrum bands with the help of dynamic spectrum access. A primary user can access the spectrum with safety and security. Further secondary (i.e., unlicensed) users may also utilize the spectrum assigned to the primary users with no interference [6]. So, cognitive IoT resolves the problem of spectrum deficiency through shared and dynamic spectrum access.

This chapter is concerned with the brief survey of the cognitive IoT followed by the discussion of its major constituent functional phases. Some major revolutionary techniques that may be collaborated with cognitive IoT are discussed next. The next section is concerned with the major application areas of cognitive IoT. At last, some future research challenges and their solutions in the cognitive IoT environment are elaborated.

5.2   LITERATURE SURVEY

A very few number of papers are based on cognitive IoT. In Ref. [1], CIoT is presented as an amalgamation of the IoT with cognitive operations to increase the throughput and achieve smartness. In Ref. [2], the authors defined CIoT as a network in which the various interconnected things are acting as delegates with least human interference. The various objects communicate with each other, use the concept of learning, storing the knowledge and adjusting to the changes. The authors in Ref. [7] presented a cognitive management system that directs the systems to judge the contiguity between the various applications and the objects, provide the smartness by minimizing the user interference and makes sure the flexibility of the supplying of the effective service. The principle and applications of the Cognitive radio with IoT has been described in Ref. [8] that also explain the benefits of adding the cognitive configuration to IoT. The authors in Ref. [9] suggested upgrading the IoT system with the cognitive configuration. They suggested that for effective machine-to-machine (M2M) communication in cognitive mode, the insight of spectrum with energy-efficient protocol stack is mandatory. In Ref. [10], the authors proposed a Code Division Multiple Access (CDMA) cognitive IoT system with expansion of frequency. The proposed system may be used for the multiplexing as well as fading channels. In Ref. [11], the authors discussed about the classification and the various technologies of IoT. The concept of spectrum sensing and the different challenges for CIoT have also been discussed. Further, a self-reorganized IoT system was suggested by the authors in Ref. [12] in which case-study of various technologies concerned with IoT system was also discussed. The authors in Ref. [13] found a model to boost up the proper utilization of spectrum and to increase the final throughput using cognitive IoT to enhance the performance of the network by using multi-objective optimization. In Ref. [14], the authors proposed a protocol based on priority similar to channel hopping techniques for device-to-device (D2D) communication in CIoT system. The consumption of energy was not found to be affected using the channel hopping. The use of genetic algorithm for the optimization of spectral efficiency as well as throughput in cognitive radio-based IoT system for the spectrum management in 5th generation has been discussed in Ref. [15]. In Ref. [16], the authors suggested a wideband cognitive IoT system based on 5G for the awareness of spectrum.

5.3 FUNCTIONAL PHASES OF COGNITIVE IoT

The cognitive features in CIoT system may be realized by three different functional phases as shown in Figure 5.1.

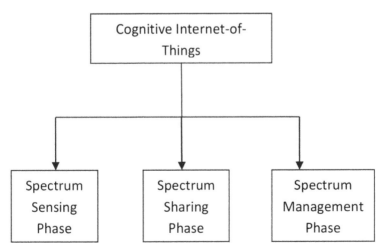

FIGURE 5.1 Functional phases for cognitive Internet of Things.

5.3.1   SPECTRUM SENSING

The spectrum phase of CIoT system is concerned with acquiring a specific spectrum sensing technique and decision-making approach [17]. There are two types of spectrum sensing:

1. **Narrowband Spectrum Sensing:** When there is appreciable information about the bandwidth and the primary user.
2. **Wideband Spectrum Sensing:** When the details of the primary user and bandwidth are not properly known to the CIoT system. Although, it utilizes the spectrum more efficiently but this approach is more complex.

Many narrowband and wideband spectrum sensing techniques have been implemented in the past [17]. The aim of decision making is to make the decision of sensing for the observation. The various factors like noise, shadowing, and fading affect the decision making [18]. The cooperation among the various devices by sharing their observations of sensing can compensate for the effect of these undesired factors.

5.3.2   SPECTRUM SHARING

After sensing, the main objective of this phase is to assign the unoccupied primary user band and then make use of that and exploit that efficiently. Spectrum sharing can be carried out with medium access control (MAC) protocol and spectrum routing. There are some criteria for the spectrum sharing [17]. The main criteria are:

- A secondary user can utilize only one assigned channel;
- Overall interference due to the secondary users must be the maximum permitted limit at the most;
- The assigned channel must complement the power and bandwidth requirements of secondary user.

5.3.3   SPECTRUM MANAGEMENT

The data exchanged in the CIoT system may contain some critical information that should be secure and private. So, secure, and private protocols and mechanisms requiring the least communication and minimum overhead must be used against the threats. The protocols to be used mainly depend upon the applications and specifications of the device [19]. The energy efficient protocols are best suited to all these requirements. The conventional security protocols and algorithms are not suitable for this purpose as most of the devices used in CIoT system are having low energy [20].

5.4 TECHNOLOGIES INTEGRABLE WITH COGNITIVE IoT

The various state-of-the-art technologies that may be used with the cognitive IoT are as follows.

5.4.1 *OPTIMIZATION SCHEMES*

In CIoT system, the different sensor nodes consume a lot of energy. Many techniques have been adopted so that this consumption of energy can be reduced by collaborative spectrum sensing [21]. But lowering the energy consumption results in the poor detection and hence system becomes less reliable [22]. So, there must be a compromise between consumption of energy and detection precision. This may be achieved by modeling in the form of multi-objective optimization.

The optimization should aim towards the minimization of energy and maximization of detection precision. One of the optimization techniques used for this purpose is Artificial Physics Optimization which is a very smart algorithm [23]. The solution of the problem is mapped to the particle having mass, position, and velocity with the various operators like initialization, encoding, and fitness function. The proposed algorithm has a very good decision-making capability as far as the detection of spectrum is concerned [24].

Further, the cognitive throughput in CIoT can be optimized by minimizing the time required for the sensing of spectrum and minimizing the packet error rate [25].

5.4.2 *BLOCKCHAIN*

It is a huge and safe data store of the individual records (i.e., blocks). The various blocks are linked to each other. In blockchain, it is immune against the data modification and hence system is resistant to hacking and cyber-crime. It is managed by a large number of users [26]. As it is owned by none, so it eliminates the failure of system.

The blockchain can be united with CIoT systems. It can support the distribution of the various services within the various devices in IoT. Cognitive IoT system with blockchain can be used for the transaction, supply chain, shipping, etc. [27]. Further, blockchain acts as a foundation for the industrial applications of CIoT [28]. Using the blockchain in IoT and cognitive systems helps in security and privacy of data [29]. Applying the blockchain in CIoT gives the spectrum access facility with absolutely no collision. It has been found to be even better than the Medium Access Control (MAC) protocol in some situations [30]. Further, by combining artificial intelligence and blockchain in CIoT systems, the various issues like randomness and complexity may be resolved.

5.4.3 MACHINE LEARNING

Machine learning (ML) techniques are the applications of Artificial Intelligence (AI) to build the statistical models in which decisions are made based on the training. So, an ML system helps to upgrade the system by learning (supervised or unsupervised) without any human interference. So, ML techniques can be very helpful in data analysis, feature extraction, etc. They provide the privacy and security in big data systems [17].

The machine learning approaches may be collaborated with cognitive IoT system as they are based on implicit learning and decision optimization and hence can improve the processing capabilities in a Cognitive IoT system based on iterative feedback. In cognitive ML, machine works like human thinking process and human intelligence. The various ML techniques like Logistic Regression (LR), Support Vector Machine (SVM), K-Nearest Neighbor (KNN), hidden Markov are most commonly used. There should be some criteria on the basis of which the most suitable ML technique may be chosen for using CIoT in the applications like health monitoring, weather forecasting, 5G networks, traffic management, decentralized networks, speech recognition, etc. [31–34].

5.4.4 *ORTHOGONAL FREQUENCY DIVISION MULTIPLEXING*

The cognitive IoT may be used in conjunction with Orthogonal Frequency Division Multiplexing (OFDM) for the purpose of spectrum sharing. OFDM is a prime technology for the IoT networks [35, 36] as OFDM has very high bandwidth and hence high data rate in the fading environment.

An OFDM based CIoT network is having a primary user P (having both transmitter and receiver) and some unlicensed IoT devices I_i (having both transmitter and receiver) where i = 1, 2,... M (say). This network model is shown in Figure 5.2. The signals in both P and I_i are modulated by OFDM. P works on some bands that are licensed. But it may face channel fading between the link connecting transmitter and receiver. So, it solicits the assistance from its adjacent I_i for the advancement of the signal to the receiving end that needs I_i to have spectrum sharing. If I_i deny to switch the signal of P because of some cost involved, P may offer some cost to I_i for the spectrum sharing. But P does not know which I_i to choose (as it does not know which channel between each transmitter of I_i and receiver of P is better in terms of least fading). There are various algorithms that have been proposed to choose I_i [37].

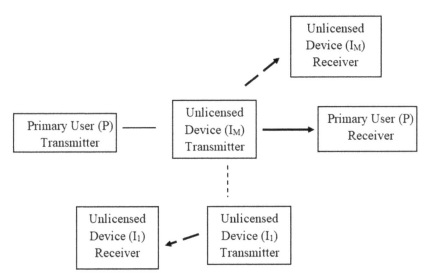

FIGURE 5.2 A CIoT model based on OFDM.

The interaction between P and I_i has three steps: Firstly, P multicasts its signal from transmitter to its own receiver as well as the transmitters of I_i that are able to decode. In the next step, the transmitters of I_i transmit that signal received to the receiver of P using the space-time coding. In the last step, the different I_i convey the signals of their own to their own receivers with different carrier.

5.5   APPLICATIONS

Many of the applications of CIoT have already turned up and some more are yet to come because of the harmony between customers, production, and internet [38]. This collaboration helps to link the various smart things that generate and send the practical data. The various applications of the CIoT system are concerned with the various technologies like mobile communication, Internet of Things, and cloud computing.

5.5.1   SENSOR NETWORKS

The cognitive IoT networks are employed for the evolution of smart or intelligent networks that are not restricted for the communication between two endpoints inside the network. It means that the IoT sensor networks are concerned with huge data and supporting details than just acting as communication link between two points. The sensor architecture is aimed towards applying them in smart cities and keeping track of many outdoor activities. This may be attained by making use of the learning technique that relates the effect of the earlier actions/decisions on the current (or future) conduct of the network. So, learning is a significant element for the cognition in an IoT network as it increases the management and hence overall performance of the network [39].

A cognitive sensor network has various cognitive components that are used for training, analyzing, and reproducing knowledge [40] as shown in Figure 5.3.

The training is based upon the long-term learning from the past action based on the feedback and behaving well in future to make the positive decisions and hence to raise the quality of information that enhances the life of network [41]. The analysis is based on responding based on the current behavior. The knowledge reproduction is concerned with the conversion of data to fruitful information.

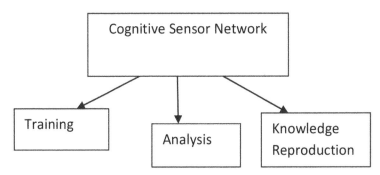

FIGURE 5.3 Various components of a cognitive sensor network.

5.5.2 HEALTH MONITORING

Cognitive Internet of Things has made appreciable progress in examining the health of the people with the help of environmental sensors and wearable sensors [42]. The environmental sensors refer to the combination of various objects connected to each other that may be employed for monitoring and measuring the various variables like humidity, temperature, dust, air quality, pressure, etc., while the wearable sensors refer to the various kinds of intelligent devices such as smartphones, wristwatches, goggles, etc., to observe the activities such as heart rate and blood pressure.

The first priority of the environmental sensors is not the monitoring of a person's health [43]. Rather, such kind of sensors can generate the information regarding health by collecting data followed by its comparison from some other sources. The data provided by environment sensor and wearable sensor is beneficial for the hospitals, healthcare agencies, various medical organizations and also for the self-health-monitoring [44]. The benefit of using wearable sensor is that the data may be collected for a much longer duration.

The data may be deposited and analyzed with any type of machine learning technique [45]. The progress in big data, data analytics and artificial intelligence techniques has made the data collection, data processing and decision-making from the various sensors quite precise and genuine [46].

Extremely care should be taken while collecting the data from these sensors as the electromagnetic radiations emitted from the cellular devices may disrupt these sensors resulting in harming the medical data [47]. So, some specific distance is to be set up to avoid the damage of these sensors due to electromagnetic interference (EMI).

5.5.3   ENERGY HARVESTING

Energy harvesting (EH) [48] based upon RF energy is a fast-growing technology. In a cognitive IoT network, the various relays pass on the required information to the destination node. In some situations, some of the relay nodes are having bound energy that deteriorates the behavior of the network. RF energy harvesting is a remedy to this constraint as it carries information along with energy [49]. It is not genuine to harvest energy from the renewable sources because of the changing weather conditions [50, 51]. There are two common methods of having this RF energy harvesting. The first method is based on time division in which the relay utilizes the power of source node in two slots-one for harvesting energy and second one for the processing of information. An alternative technique is to divide the source node power by the relay into two parts one part for the purpose of EH and other part for the decoding of information. Although the first method is more beneficial because of utilizing full energy in each time slot, still it is least commonly used because it requires a very precise synchronization between the various nodes.

IoT system must have the cognitive features as the cognitive radio has the capability to use the unexploited spectrum smartly (i.e., they are self-configurable) by not overlapping with the other users [52–54], thus solving the problem of spectrum deficiency. This is possible because cognitive IoT system is capable of sensing and exploiting the frequency spectrum in the neighboring areas. In other words, spectrum leasing is required by cognitive IoT for assisting a broad range of applications. The unlicensed (or secondary) users may access the spectrum assigned to licensed (or primary) users provided no interference exists [55]. This is simply because more than 70% of the spectrum allocated to licensed users is not properly utilized even in the congested regions [56]. For example, one of the major features of the cognitive IoT is to explore the spectrum for the unutilized Television White Spaces (TVWS) that may be employed for the EH applications to use for the channeling of data. The decision related to transmission is made if the harvested energy is enough to be radiated. Any algorithm used for the EH must take into account the optimization of factors like cost parameters, quality of data and time delay [57].

5.5.4   NEURAL NETWORKS

The conventional techniques of neural networks do not fit into the applications of IoT as these techniques need huge and extraordinary computation like Graphics Processing Unit. A neural network generally uses the machine learning algorithms for the mathematical modeling using neurons. To run these complex and high standard machine learning algorithms on Internet-of-Thing's system, it requires well-organized and effective resources with large power of processing. The data compression is a major factor to run the deep learning algorithms on IoT system having limited resources. This data compression changes highly dense neural network to sporadic neural network that reduces the computational burden by a large factor and hence the network can easily be run on IoT platform. Further, IoT devices in many cases generate diversified and inhomogeneous data, whose transmission in many situations may result in the noise arising in the data that may have a negative impact on the final results. So, our solution must be such that it can withstand noise. Moreover, hardware used should be special-purpose to minimize the dissipation of power by the neural networks [58–60].

By adopting the above methodology, the deep learning based neural techniques may easily be run on IoT platform.

5.5.5   DATA ANALYTICS

The devices at the fringe of Internet-of-Thing's platform have micro-operating system installed. An application system may be installed that is able to perform the analytics on the data that helps to select the required data for storing and deriving predictions. In other words, it is all about doing the analysis in real-time of the data generated. For example, analysis of real time video through CCTV cameras works out better than sending that video on IoT system first and to analyze it later. But there are some drawbacks in data analytics. Only small amount of data is available for the processing followed by the analysis that is transmitted. In other words, some of the data is not considered for the analysis. In some situations, the data are so crucial that it must not be neglected. For example, in railway systems, the entire data need to be channeled for the analysis that can help to detect the flaw in the functioning of the engine.

An example of such type of application system is Node-Red that works as a programming tool for IoT platform. It may be installed on Raspberry Pi and many other devices [58, 61]. IoT in data analytics is useful in manufacturing, health systems and telematics, etc.

5.5.6   5G NETWORKS

5G or 5th Generation technology [62, 63] is considered as a major factor in the evolution of Cognitive IoT and its applications. 5G may be regarded as a superset of 4G, as shown in Figure 5.4. 5G has all the attributes as 4G like data, voice, connectivity but is better in all respects than 4G.

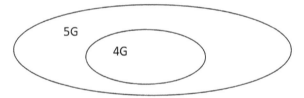

FIGURE 5.4   4G is shown as a subset of 5G.

5G is also a superset of WiMAX, Wi-Fi, and many more wireless technologies. 5G mobile network is likely to be inhomogeneous as far as devices, technologies, and cell size are concerned. The technology used may be based on either of full-duplex, MIMO-OFDM, and non-orthogonal multiple access (NOMA) [64]. Cell size may be of the order of either macro cells or microcells or even pico-cells or femtocells [65]. It exhibits higher coverage, massive capacity, tremendous connectivity, higher throughput, higher spectral efficiency, much improved energy efficiency, advanced privacy, and much higher speed (of the order of 10 Gbps or even higher than that) with quite less latency time [66]. Therefore, there is a massive demand for the frequency spectrum that needs the utilization and proper management [67]. This demand may be fulfilled by dynamic approach (or shared approach) of spectrum that is concerned with the utilization of the underutilized bands for the secondary users [66, 68]. 5G is likely to be the very first platform that would assist billions of devices with respect to each other. Device-to-device (D2D) communication is the emerging technology in 5G networks that helps to manage the shortest or direct connection between the users in case they are at a large distance from the base station [69]. But with the increase in the number of devices, chances of interference are also quite high. So, these networks will be required to have self-organization to have the optimized system so that they can sense which portion of the spectrum should be used for the creation of the wireless connection.

5.5.7 INDUSTRIAL APPLICATIONS

Cognitive IoT may be used in various industrial applications. For example, it may be used:

- to enhance the quality of service in the various smart grid applications [70];
- to reduce the time delay for routing in the industrial applications in which time is a very crucial factor [71];
- to realize the hand-off in the spectrum of INTERNET-of-things [72];
- to protect the end-points in the cyber-physical systems by the switching of frequency at the endpoints [73].

Based on the above discussions, it may be concluded that cognitive IoT may be used in useful applications like Smart Sensors, Smart Health, Smart Energy, Smart Homes, Smart Parking, Smart Grids, Smart Tags, etc. [63].

5.6 CHALLENGES AND SOLUTIONS IN COGNITIVE IoT

There are a lot of potential challenges or issues in cognitive IoT system. Here, nine major challenges have been identified (as shown in Figure 5.5) -based on the research in this field. These challenges are described in subsections.

5.6.1 TECHNICAL CHALLENGES

The deployment of applications of cognitive Internet of Things system is very tough because of its heterogeneous structure, scalability, and limited resources. A multidimensional service platform is a good solution to resolve this issue [74].

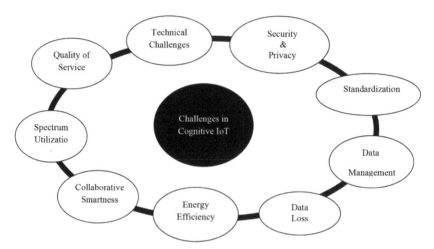

FIGURE 5.5   Major challenges in cognitive IoT.

5.6.2   SECURITY AND PRIVACY ISSUES

With high system capacity, high data rate and heterogeneity in the CIoT network, security, and privacy are big issues. So, the system should have a great potential at the device level as well as network level to resolve the various issues related to security and privacy, especially in the important applications like health monitoring, smart city, etc. [75, 76].

5.6.3   STANDARDIZATION ISSUE

Standardization is a major step for the growth of Cognitive IoT networks. Due to the assorted class of the different devices and networks, there is non-existence of standardization with regards to the protocols, security, privacy, cryptography; data aggregation, etc., as far as Cognitive IoT and its applications are concerned. Investing IoT in the form of a service and refining the structure of Cognitive IoT may lead to its standardization in the upcoming years [76, 77].

5.6.4 DATA MANAGEMENT ISSUES

With the more and more use of Cognitive IoT, higher amount of data is being generated that also includes the data created from the various sensor networks. This data may be in organized or unorganized form. The main challenge lies in the use of the various mining techniques to mine the data from the sensor's networks. The solution lies in the implementation of the various data mining methods of the advanced level to mine the data from the sensor networks [78]. Also, a smart sensing system may be made that can imitate the cognition among the human beings to link the various types of data by learning, determining, and concluding.

5.6.5 DATA LOSS ISSUE

In the case of data assembled by a microcontroller and transferred to a smart device, there may be a case of data loss due to disconnection. This is a huge challenge, especially in the case of health management. The solution is to save the temporary data in a microcontroller having huge memory space [47].

5.6.6 ENERGY EFFICIENCY ISSUE

All the sensors, things, and smart devices used in a cognitive IoT system are less energy-efficient and hence have fewer lifetimes in many cases. This issue may be solved to a larger extent by using the reliable devices having appreciable processing power and considerable energy efficiency [47]. When a large number of sensors are used for sensing the spectrum, energy consumed is more although the detection is precise in this case. The balance between energy efficiency and detection perfection may be achieved through some decision-making optimization technique [24].

5.6.7 COLLABORATIVE SMARTNESS

As IoT is concerned with machine to machine (M2M), device to device (D2D) and human to machine (H2M) communication, so the devices must have the interactive capabilities. The cooperation based on the sensing among the various smart and intelligent devices in cognitive IoT system is also an issue in various circumstances because of the sensing precision and sensing delay. This issue may be addressed to make the task of decision-making and data processing quite easier by using feature-based detection and matched filter. Besides, the compressed form of sensing may also be used to control the sampling frequency in a large range of spectrum sensing [47, 79].

5.6.8   SPECTRUM UTILIZATION

The major inspiration for using cognitive IoT system is to optimize the spectral efficiency. Most of the cognitive IoT systems are not systematic in the proper utilization of spectrum. To address this challenge, a model based on the efficient spectrum access may be used for the various dynamic applications of Cognitive IoT [80]. One common example of the efficient spectrum utilization model is dynamic spectrum access (DSA) model. In many situations, there occurs collision in the network because of the multi-user system. Models such as fair resource allocation model [81] and priority access model [82] are useful in such cases. Further, one more model called energy-efficient spectrum access (EESA) has been suggested in Ref. [80] that resolves the interference and also increases the throughput.

5.6.9   QUALITY OF SERVICE (QOS)

QoS is a major factor to determine the level up to which the service expectations are fulfilled. The level of interaction between machines and humans also determines QoS. With the technologies such as virtual reality, graphical models have evolved into the natural models that have increased QoS to some extent. Still, a next generation model is yet to be made that can judge the sentiment level of the users to yield more natural service. Further, various techniques based on the fading, inactive time, and channel usage, etc., have also been proposed to improve the quality of service [83, 84].

5.7   CONCLUSION

The concept of Cognitive IoT has been discussed in detail in this article. A recent survey related to the Cognitive IoT is discussed first. The three constituent functional phases of the cognitive IoT are discussed in brief. The various thrust areas like optimization schemes, blockchain, machine learning and Orthogonal Frequency Division Multiplexing (OFDM) are also discussed in collaboration with cognitive IoT. Further, the various important applications of Cognitive IoT in sensor networks, health monitoring, energy harvesting, neural networks, data analytics, 5G Networks and industries are also discussed. The various research challenges with their solutions in the field of Cognitive IoT are discussed in the last section.

KEYWORDS

- **applications**
- **challenges**
- **cognitive Internet of Things**
- **Internet of Things**
- **phases**
- **spectrum**
- **technologies**

REFERENCES

1. Zhang, M., Zhao, H., Zheng, R., Wu, Q., & Wei, W., (2012). Cognitive Internet of Things: Concepts and application example. *International Journal of Computer Science Issues (IJCSI), 9*(6), 151.
2. Wu, Q., Ding, G., Xu, Y., Feng, S., Du, Z., Wang, J., & Long, K., (2014). Cognitive Internet of Things: A new paradigm beyond connection. *IEEE Internet of Things Journal, 1*(2), 129–143.
3. https://www.ibmbigdatahub.com/blog/what-cognitive-iot (accessed on 16 November 2021).
4. Zhang, M., Qiu, Y., Zheng, R., Bai, X., Wei, W., & Wu, Q., (2015). A novel architecture for cognitive Internet of Things. *International Journal of Security and its Applications, 9*(9), 235–252.
5. Alhussein, M., Muhammad, G., Hossain, M. S., & Amin, S. U., (2018). Cognitive IoT-cloud integration for smart healthcare: Case study for epileptic seizure detection and monitoring. *Mobile Networks and Applications, 23*(6), 1624–1635.

6. Elderini, T., Kaabouch, N., & Reyes, H., (2017). Channel quality estimation metrics in cognitive radio networks: A survey. *IET Communications, 11*(8), 1173–1175.

7. Vlacheas, P., Giaffreda, R., Stavroulaki, V., Kelaidonis, D., Foteinos, V., Poulios, G., & Moessner, K., (2013). *Enabling Smart Cities Through a Cognitive Management Framework for the Internet of Things, 51*(6), 102–111. IEEE Communications Magazine.

8. Shah, M. A., Zhang, S., & Maple, C., (2013). Cognitive radio networks for Internet of Things: Applications, challenges, and future. In: *Proceedings of 19th International Conference on Automation and Computing* (pp. 1–6). London, UK, IEEE.

9. Aijaz, A., & Aghvami, A. H., (2015). Cognitive machine-to-machine communications for Internet of Things: A protocol stack perspective. *IEEE Internet of Things Journal, 2*(2), 103–112.

10. Hu, S., Guo, H., Jin, C., Huang, Y., Yu, B., & Li, S., (2016). Frequency-domain oversampling for cognitive CDMA systems: Enabling robust and massive multiple access for Internet of Things. *IEEE Access, 4*, 4583–4585.

11. Khan, A. A., Rehmani, M. H., & Rachedi, A., (2016). When cognitive radio meets the Internet of Things? In: *Proceedings of 2016 International Wireless Communications and Mobile Computing Conference (IWCMC 2016)* (pp. 469–474). Paphos, Cyprus, IEEE. doi: 10.1109/IWCMC.2016.7577103.

12. Khan, A. A., Rehmani, M. H., & Rachedi, A., (2017). Cognitive-radio-based Internet of Things: Applications, architectures, spectrum related functionalities, and future research directions. *IEEE Wireless Communications, 24*(3), 17–25.

13. Han, R., Gao, Y., Wu, C., & Lu, D., (2018). An effective multi-objective optimization algorithm for spectrum allocations in the cognitive-radio-based Internet of Things. *IEEE Access, 6*, 12858–12867.

14. Liu, X., & Xie, J. L., (2017). Priority-based spectrum access in cognitive D2D networks for IoT. In: *2017 IEEE International Conference on Communications (ICC-2017)* (pp. 1–6). Paris, France. IEEE. doi: 10.1109/ICC.2017.7996716.

15. Tseng, F. H., Chao, H. C., & Wang, J., (2015). Ultra-dense small cell planning using cognitive radio network toward 5G. *IEEE Wireless Communications., 22*(6), 76–83.

16. Liu, X., He, D., & Jia, M., (2017). 5G-based wideband cognitive radio system design with cooperative spectrum sensing. *Physical Communication, 25*, 539–545.

17. Awin, F. A., Alginahi, Y. M., Abdel-Raheem, E., & Tepe, K., (2019). Technical issues on cognitive radio-based Internet of Things systems: A survey. *IEEE Access, 7*, 97887–97908.

18. Al Hussien, N., (2014). *Narrowband and Wideband Spectrum Sensing for Cognitive Radio Networks in a Log-Normal Shadowing Environment*, Ph. D. dissertation, Dept. Elect. Eng., United Arab Emirates Univ., United Arab Emirates.

19. Hossain, M. M., Fotouhi, M., & Hasan, R., (2015). Towards an analysis of security issues, challenges, and open problems in the Internet of Things. In: *2015 IEEE World Congress on Services* (pp. 21–28). New York, USA. IEEE. doi: 10.1109/SERVICES.2015.12.

20. Sadeghi, A. R., Wachsmann, C., & Waidner, M., (2015). Security and privacy challenges in industrial Internet of Things. In: *2015 52nd ACM/EDAC/IEEE Design Automation Conference (DAC)* (pp. 1–6). San Francisco USA. IEEE. doi: 10.1145/2744765.2747942.

21. Cichoń, K., Kliks, A., & Bogucka, H., (2016). Energy-efficient cooperative spectrum sensing: A survey. *IEEE Communications Surveys & Tutorials, 18*(3), 1861–1886.
22. Chai, Z. Y., Wang, B., & Li, Y. L., (2014). Spectrum allocation of cognitive radio network based on artificial physics optimization. *Acta Physica Sinica, 63*(22), 228802.
23. Zhong, M., Zhang, H., & Ma, B., (2016). APO-based parallel algorithm of channel allocation for cognitive networks. *China Communications, 13*(6), 100–105.
24. Li, Y., Wang, H., & Chai, Z., (2019). Multi-objective optimization of spectrum detection in cognitive IoT using artificial physics. *Journal of the Chinese Institute of Engineers, 42*(3), 219–224.
25. Zhang, L., & Liang, Y. C., (2019). Joint spectrum sensing and packet error rate optimization in cognitive IoT. *IEEE Internet of Things Journal, 6*(5), 7816–7827.
26. Benchoufi, M., & Ravaud, P., (2017). Blockchain technology for improving clinical research quality. *Trials, 18*(1), 1–5.
27. Brody, P., & Pureswaran, V., (2014). *Device Democracy: SAVING the Future of the Internet of Things.* IBM.
28. Bahga, A., & Madisetti, V. K., (2016). Blockchain platform for industrial Internet of Things. *Journal of Software Engineering and Applications, 9*(10), 533–546.
29. Saghiri, A. M., Vahdati, M., Gholizadeh, K., Meybodi, M. R., Dehghan, M., & Rashidi, H., (2018). A framework for cognitive Internet of Things based on blockchain. In: *2018 4*th *International Conference on Web Research (ICWR)* (pp. 138–143). Tehran, Iran. IEEE. doi: 10.1109/ICWR.2018.8387250.
30. Kotobi, K., & Bilén, S. G., (2017). Blockchain-enabled spectrum access in cognitive radio networks. In: *2017 Wireless Telecommunications Symposium (WTS)* (pp. 1–6). Chicago, USA. IEEE. doi: 10.1109/WTS.2017.7943523.
31. Mahdavinejad, M. S., Rezvan, M., Barekatain, M., Adibi, P., Barnaghi, P., & Sheth, A. P., (2018). Machine learning for Internet of Things data analysis: A survey. *Digital Communications and Networks, 4*(3), 161–175.
32. Kumar, P. M., Gandhi, U. D., A novel three-tier Internet of Things architecture with machine learning algorithm for early detection of heart diseases. *Computers & Electrical Engineering, 65, 222–235.*
33. Bkassiny, M., (2012). Li, Y., Jayaweera, S. K., A survey on machine-learning techniques in cognitive radios. *IEEE Communications Surveys & Tutorials, 15*(3), 1136–1155.
34. Xu, Y., (2015). Recent machine learning applications to Internet of Things (IoT). *Recent Advances in Networking.* [online] Available at: https://www.cs.wustl.edu/~jain/cse570-15/ftp/iot_ml/ (accessed on 16 November 2021).
35. Lu, W., & Wang, J., (2014). Opportunistic spectrum sharing based on full-duplex cooperative OFDM relaying. *IEEE Communications Letters, 18*(2), 241–244.
36. Zhang, L., Ijaz, A., Xiao, P., Molu, M. M., & Tafazolli, R., (2017). Filtered OFDM systems, algorithms, and performance analysis for 5G and beyond. *IEEE Transactions on Communications, 66*(3), 1205–1218.
37. Lu, W., Hu, S., Liu, X., He, C., & Gong, Y., (2019). Incentive mechanism based cooperative spectrum sharing for OFDM cognitive IoT network. *IEEE Transactions on Network Science and Engineering, 7*(2), 662–672.
38. Vermesan, O., & Friess, P., (2014). River publishers' series in communication. *Internet of Things-from Research and Innovation to Market Deployment* (Vol. 29). River publishers: Aalborg.

39. Al-Turjman, F., & Gunay, M., (2016). CAR Approach for the Internet of Things. *Canadian Journal of Electrical and Computer Engineering, 39*(1), 11–18.

40. Al-Turjman, F. M., (2017). Information-centric sensor networks for cognitive IoT: An overview. *Annals of Telecommunications, 72*(1, 2), 3–18.

41. Bala, I., Bhamrah, M. S., & Singh, G., (2017). Rate and power optimization under received-power constraints for opportunistic spectrum-sharing communication. *Wireless Personal Communication, 96,* 5667–5685.

42. Bala, I., Bhamrah, M. S., & Singh, G., (2017). Capacity in fading environment based on soft sensing information under spectrum sharing constraints. *Wireless Network, 23,* 519–531.

43. Noury, N., Picard, R., Billebot, M. N., Durand-Salmon, F., Lewkowicz, M., & Noat, H., (2018). Challenges and limitations of data capture versus data entry. In: *Connected Healthcare for the Citizen* (pp. 85–97). Robert Picard; Elsevier. doi: https://doi. org/10.1016/B978-1-78548-298-4.50007-5.

44. Cao, H., Leung, V., Chow, C., & Chan, H., (2009). *Enabling Technologies for Wireless Body Area Networks: A Survey and Outlook, 47*(12), 84–93. IEEE Communications Magazine.

45. Bala, I., Bhamrah, M. S., & Singh, G., (2019). Investigation on outage capacity of spectrum sharing system using CSI and SSI under received power constraints. *Wireless Network, 25,* 1047–1056.

46. Pateraki, M., Fysarakis, K., Sakkalis, V., Spanoudakis, G., Varlamis, I., Maniadakis, M., & Loutsetis, E., (2020). Biosensors and Internet of Things in smart healthcare applications: Challenges and opportunities. In: Dey, N., Ashour, A. S., & Bhatt, C., (eds.), *Wearable*, and *Implantable Medical Devices* (Vol. 7, pp. 25–53). Academic Press. doi: https://doi.org/10.1016/c2017-0-03249-4.

47. Aileni, R. M., Suciu, G., Suciu, V., Pasca, S., & Strungaru, R., (2019). Health monitoring using wearable technologies and cognitive radio for IoT. In: Rehmani, M. H., & Dhaou, R., (eds.), *Cognitive Radio, Mobile Communications and Wireless Networks* (pp. 143–165). Springer: Cham. doi: https://doi.org/10.1007/978-3-319-91002-4_6.

48. Bouabdellah, M., El Bouanani, F., Sofotasios, P. C., Muhaidat, S., Da Costa, D. B., Mezher, K., & Karagiannidis, G. K., (2019). Cooperative energy harvesting cognitive radio networks with spectrum sharing and security constraints. *IEEE Access, 7,* 173329–173343.

49. Bhowmick, A., Yadav, K., Roy, S. D., & Kundu, S., (2017). Throughput of an energy harvesting cognitive radio network based on prediction of primary user. *IEEE Transactions on Vehicular Technology, 66*(9), 8119–8128.

50. Mahmoud, H. H., ElAttar, H. M., Saafan, A., & ElBadawy, H., (2017). Optimal operational parameters for 5G energy harvesting cognitive wireless sensor networks. *IETE Technical Review, 34*(1), 62–72.

51. Liu, X., Li, F., & Na, Z., (2017). Optimal resource allocation in simultaneous cooperative spectrum sensing and energy harvesting for multichannel cognitive radio. *IEEE Access, 5,* 3801–3812.

52. Tragos, E. Z., & Angelakis, V., (2013). Cognitive radio inspired M2M communications. In: *2013 16ᵗʰ International Symposium on Wireless Personal Multimedia Communications (WPMC)* (pp. 1–5). Atlantic City, USA, IEEE.

53. Cordeiro, C., Challapali, K., Birru, D., & Shankar, S., (2006). IEEE 802.22: An introduction to the first wireless standard based on cognitive radios. *Journal of Communications, 1*(1), 38–47.

54. Ozger, M., Cetinkaya, O., & Akan, O. B., (2018). Energy harvesting cognitive radio networking for IoT-enabled smart grid. *Mobile Networks and Applications, 23*(4), 956–966.

55. Mishra, N., Kundu, S., Mondal, S., & Roy, S. D., (2019). Cognitive machine to machine communication with energy harvesting in IoT networks. In: *2019 11ᵗʰ International Conference on Communication Systems & Networks (COMSNETS)* (pp. 672–677). Bengaluru, India, IEEE.

56. Qu, Z., Xu, Y., & Yin, S., (2014). A novel clustering-based spectrum sensing in cognitive radio wireless sensor networks. In: *2014 IEEE 3ʳᵈ International Conference on Cloud Computing and Intelligence Systems* (pp. 695–695). Shenzhen, China, IEEE.

57. Lad, S. P., Kulkarni, V. P., & Joshi, R. D., (2019). An online algorithm for energy harvesting cognitive radio IoT network. In: *2019 IEEE International Conference on Advanced Networks and Telecommunications Systems (ANTS)* (pp. 1–6). Goa, India, IEEE.

58. Chandra, N., Khatri, S. K., & Som, S., (2019). Business models leveraging IoT and cognitive computing. In: *2019 Amity International Conference on Artificial Intelligence (AICAI)* (pp. 796–800). Dubai, United Arab Emirates, IEEE.

59. Du, R., Magnusson, S., & Fischione, C., (2020). The Internet of Things as a deep neural network. In: *IEEE Communications Magazine* (Vol. 58, No. 9, pp. 20–25). doi: 10.1109/MCOM.001.2000015.

60. Razafimandimby, C., Loscri, V., & Vegni, A. M., (2016). A neural network and IoT based scheme for performance assessment in internet of robotic things. In: *2016 IEEE first International Conference on Internet of Things Design and Implementation (IoTDI)* (pp. 241–246). Berlin, Germany, IEEE.

61. Tulasi, B., & Girish, J., (2016). Blending IoT and big data analytics. *International Journal of Engineering Sciences & Research Technology, 5*(4), 192–196.

62. Intelligence, G. S. M. A., (2014). *Understanding 5G: Perspectives on Future Technological Advancements in Mobile* (pp. 1–26). White paper.

63. Katzis, K., & Ahmadi, H., (2016). Challenges implementing Internet of Things (IoT) using cognitive radio capabilities in 5G mobile networks. In: Mavromoustakis, C. X., Mastorakis, G., & Batalla, J. M., (eds.), *Internet of Things (IoT) in 5G Mobile Technologies* (pp. 55–76). Springer: Cham.

64. Zhang, X., & Wang, J., (2017). Heterogeneous QoS-driven resource allocation over MIMO-OFDMA based 5G cognitive radio networks. In: *2017 IEEE Wireless Communications and Networking Conference (WCNC)* (pp. 1–6). San Francisco, USA, IEEE.

65. Park, H., & Hwang, T., (2016). Energy-efficient power control of cognitive femto users for 5G communications. *IEEE Journal on Selected Areas in Communications, 34*(4), 772–785.

66. Madan, H. T., & Basarkod, P. I., (2018). A survey on efficient spectrum utilization for future wireless networks using cognitive radio approach. In: *2018 4ᵗʰ International Conference on Applied and Theoretical Computing and Communication Technology (iCATccT)* (pp. 47–53). Mangalore, India, IEEE.

67. Franco, C. A. S., De Marca, J. R. B., & Siqueira, G. L., (2016). A cognitive and cooperative SON framework for 5G mobile radio access networks. In: *2016 IEEE Globecom Workshops (GC Wkshps)* (pp. 1–6). Washington DC, USA, IEEE.

68. Sahoo, P. K., Mohapatra, S., & Sheu, J. P., (2017). Dynamic spectrum allocation algorithms for industrial cognitive radio networks. *IEEE Transactions on Industrial Informatics, 14*(7), 3031–3043.

69. Tsiropoulos, G. I., Yadav, A., Zeng, M., & Dobre, O. A., (2017). Cooperation in 5G HetNets: Advanced spectrum access and D2D assisted communications. *IEEE Wireless Communications, 24*(5), 110–117.

70. Shah, G. A., Gungor, V. C., & Akan, O. B., (2013). A cross-layer QoS-aware communication framework in cognitive radio sensor networks for smart grid applications. *IEEE Transactions on Industrial Informatics, 9*(3), 1477–1485.

71. Tang, F., Tang, C., Yang, Y., Yang, L. T., Zhou, T., Li, J., & Guo, M., (2016). Delay-minimized routing in mobile cognitive networks for time-critical applications. *IEEE Transactions on Industrial Informatics, 13*(3), 1398–1405.

72. Oyewobi, S. S., Hancke, G. P., Abu-Mahfouz, A. M., & Onumanyi, A. J., (2019). An effective spectrum handoff based on reinforcement learning for target channel selection in the industrial Internet of Things. *Sensors, 19*(6), 1395.

73. Cheng, B., Zhang, J., Hancke, G. P., Karnouskos, S., & Colombo, A. W., (2018). *Industrial Cyber-Physical Systems: Realizing Cloud-Based Big Data Infrastructures, 12*(1), 25–35. IEEE Industrial Electronics Magazine.

74. Zhao, S., Yu, L., & Cheng, B., (2016). An event-driven service provisioning mechanism for IoT (Internet of Things) system interaction. *IEEE Access, 4*, 5038–5051.

75. Elkhodr, M., Shahrestani, S., & Cheung, H., (2016). *The Internet of Things: New Interoperability, Management, and Security Challenges.* arXiv preprint arXiv: 1604.04824.

76. Li, F., Lam, K. Y., Li, X., Sheng, Z., Hua, J., & Wang, L., (2019). Advances and emerging challenges in cognitive Internet of Things. *IEEE Transactions on Industrial Informatics, 16*(8), 5489–5496.

77. Banafa, A., (2016). IoT standardization and implementation challenges. *IEEE Internet of Things Newsletter, 2016,* 1–10.

78. Lee, I., & Lee, K., (2015). The Internet of Things (IoT): Applications, investments, and challenges for enterprises. *Business Horizons, 58*(4), 431–440.

79. Gao, Y., Qin, Z., Feng, Z., Zhang, Q., Holland, O., & Dohler, M., (2016). Scalable and reliable IoT enabled by dynamic spectrum management for M2M in LTE-A. *IEEE Internet of Things Journal, 3*(6), 1135–1145.

80. Fatima, R., Humera, T. S., & Khanam, R., (2019). Energy-efficient spectrum access design for cognitive radio wireless sensor network. In *2019 International Conference on Communication and Electronics Systems (ICCES)* (pp. 6–11). Coimbatore, India, IEEE.

81. Gai, Y., & Krishnamachari, B., (2011). Decentralized online learning algorithms for opportunistic spectrum access. In: *2011 IEEE Global Telecommunications Conference-GLOBECOM 2011* (pp. 1–6). Houston, USA, IEEE.

82. Anandkumar, A., Michael, N., Tang, A. K., & Swami, A., (2011). Distributed algorithms for learning and cognitive medium access with logarithmic regret. *IEEE Journal on Selected Areas in Communications, 29*(4), 731–745.

83. Zhu, J., Song, Y., Jiang, D., & Song, H., (2016). Multi-armed bandit channel access scheme with cognitive radio technology in wireless sensor networks for the Internet of Things. *IEEE Access, 4,* 4609–4617.

84. Zhang, Y., Ma, X., Zhang, J., Hossain, M. S., Muhammad, G., & Amin, S. U., (2019). Edge intelligence in the cognitive Internet of Things: Improving sensitivity and interactivity. *IEEE Network, 33*(3), 58–64.

CHAPTER 6

IoT for Underground Communication Networks: A Review

SIMARPREET KAUR

Department of Electronics and Communication Engineering, BBSBEC, Fatehgarh Sahib, Punjab, India, E-mail: rabbyshabby6@gmail.com

6.1 INTRODUCTION

Wireless underground sensor networks (WUSNs) are a specialized class of Internet of Things (IoT) and are one of the most promising research areas in recent years to extend the communication capabilities to the motes that are covered up in the ground and can connect through soil [1]. WUSNs differ from physical wireless sensor networks in the sense that the signals in terrestrial networks traverse multiple layers, i.e., soil, air, and a medium interface. But, WUSNs are a particular kind of Wireless Sensor Networks (WSNs) that mainly emphasize on the use of sensors for communication through soil, which is a very complex procedure as compared to the wireless communication through air. WUSN have various applications with numerous advantages, like ease of deployment, reliability, flexible networking, high fault-tolerance capability, concealment, timeliness of data, mobility, and coverage density, as compared to non-wireless sensors [2]. The major application areas of WUSNs include healthcare applications, emergency rescue, intruder detection, intelligent irrigation, infrastructure monitoring, sports field maintenance, border patrol, target location, environmental, and industrial monitoring, and guiding various conditions where the use of sensors above ground are not suitable like pollution control, detection of natural disasters, military applications and

Harnessing the Internet of Things (IoT) for Hyper-Connected Smart World.
Indu Bala and Kiran Ahuja (Eds.)

testing water content and soil properties. In these applications, various sensor nodes continuously collect physical data to assist application-specific processing and analysis [3]. In spite of their prospective advantages as compared with the above-ground environment, realization of WUSN is still very challenging, because of the complex and substantial impact of soil characteristics on communication. So, it is important to exploit actual soil condition statistics by deployment of an underground sensor network, which is accomplished by Internet of Underground Things (IoUT). Also, a very efficient and small size antenna to fit on the sensor surfaces is needed for wireless sensor network communication. As it is difficult to unearth and recharge the underground sensor nodes, so, the efficient communication protocols need to be designed. WUSNs being a new research area is in the experimental phase and it will take some time to get mature products available in the market [3].

6.2   BACKGROUND

There is a branch of IoT, that deals with the underground things, i.e., sensors, addresses their communication issues and their networking protocols. This branch is known as the Internet of Underground Things (IoUT). Integration of sensing and communication in WUSN sensor nodes is accomplished by IoUT [4]. A wireless sensor network is made up of thousands of small and economical motes, located statically or deployed dynamically to observe the projected surroundings. Because of the smaller size, these nodes have a number of restrictions. The basic function of a WSN is monitoring. WUSNs have numerous applications in the field of geology, commercial agriculture, security, and navigation that has encouraged substantial consideration to their ability to monitor many underground situations. Specifically, underground sensor motes are useful in agriculture to monitor the condition of soil such as mineral and moisture content [5]. The integrity of underground structures such as plumbing can also be monitored successfully by the sensors [6], and buried seismometers can monitor earthquake and landslide [7]. Recent trend of concealed sensing consists of installing a hidden sensor, like that of Figure 6.1, and to connect it to a data-logger on the surface above soil to store the evaluations of sensor which can be retrieved later. There may be a mechanism for physical or single-hop Wi-Fi connection to a central sink in a data-logger, but mainly the data-logger is physically visited to manually retrieve the data [8]. Most of the present solutions require the deployment of sensor devices at the upper layer and bound to a hidden sensor [9].

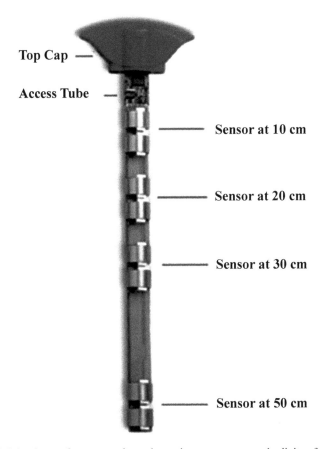

Top Cap

Access Tube

Sensor at 10 cm

Sensor at 20 cm

Sensor at 30 cm

Sensor at 50 cm

FIGURE 6.1 Sensor from measuring volumetric water content and salinity of soil at various depths.
Source: Adapted from: Ref. [8]. © www.campbellsci.com.

6.3 LIMITATIONS OF EXISTING CHANNELS

The main limitation is to make reliable and efficient concealed wireless communication possible between underground sensors. Connectivity and communication performance in IoUT based WUSNs is directly depen-dent on weather, temperature, soil moisture, soil composition, and soil homogeneity and other such factors. Indeed, atmosphere has a generous and straight impact on the communication success in underground networks. Moreover, other factors like the frequency of the EM wave and the burial depth also strongly affect the communication in wireless underground networks. Limitations of the current existing underground sensor networks can be addressed by WUSNs in the following ways [3]:

1. **Timeliness of Data:** It is possible to store data readings in data-loggers for retrieving it later. Sensor readings can be forwarded in real-time to a central sink by WUSNs.

2. **Concealment:** In present-day concealed sensing systems, motes or data-loggers are required to be extended at the ground with cabling heading towards underground sensor nodes [9, 10]. Many landscaping machinery such as tractors and lawnmowers, and agriculture can harm the devices of the sensing system which are aboveground. When sports fields or gardens are monitored, it is sometimes unacceptable to allow visible devices for performance or esthetic reasons. Alternatively, all the equipment required for transmitting and sensing is placed underground, so that it is not visible, is secured from theft or damage and is not endangered from being damaged by surface equipment.

3. **Reliability:** A data-logger may possibly have many sensor nodes linked to it and if it fails, the sensor network may stop. Sensor nodes of every data-logger may be extended over a large area, so, if the data-logger fails, it may perhaps be disastrous to the detecting application. But, in WUSNs, every sensor is able to forward interpretations independently, thus, removing the requirement of a data-logger and also the cable required to be concealed between a data-logger and a sensor node is eradicated. WUSNs are self-healing in that if there is any device failure, it can be repeatedly routed from one place to another, and the network machinist can be notified to device breakdown in real-time.

4. **Ease of Positioning:** For expanding the area of existing concealed sensing systems, it is required to deploy extra data-loggers and concealed cabling. Although the underground monitoring is done by terrestrial WSN technology as in Ref. [9], still, it is required to deploy underground wiring so that a sensor can be connected to a surface device. If it is required to deploy further sensor nodes in a WUSN, they might be located at the desired position and making

this sure that they are inside the communication area of another machine.

5. **Coverage Density:** In existing underground applications, sensors are usually deployed near their data-logger which is controlling them, so that the distance between them can be minimized. Because of this, coverage denseness can be irregular and becomes big near the data-logger and low in the environment.

Above discussion clarifies the benefits of WUSNs, but to make them feasible practically, many research demands must be focused on. It is necessary to review the existing frameworks for earthbound WSNs, together with their communication protocols and hardware because the underground environment is an unfriendly habitation for wireless communication.

6.4 CONCEALED COMMUNICATION NETWORKS CLASSIFICATION

We can classify Wireless Underground Communication Networks (WUCNs) into two main categories: Wireless Communication Networks for Mines and Tunnels and Wireless Underground Sensor Networks (WUSNs) as shown in Figure 6.2. There are many resolutions that concentrate on concealed communication in mines and/or tunnels [11, 14]. In these applications, the network is located underground, but there are voids existing below the ground, which make communication possible through air. To communicate through these voids is very challenging as compared to the communication in terrestrial WSNs; still there are similarities in the channel characteristics with that of the terrestrial WSNs.

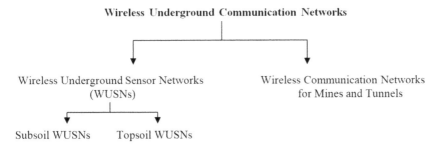

FIGURE 6.2   Wireless underground communication networks (WUCNs) classification.

WUSNs consist mainly of sensors which are hidden under the ground for making communication through soil possible. Most of the application areas of WUSNs including logical environmental monitoring, agriculture, and safety confine the underground sensor nodes in the top few meter layers of the soil. This section of the earth is generally categorized into two main sections [8]: The topsoil section denotes shallower section from either the first 30 cm of soil, or the root growth layer. The 30–100 cm section below the topsoil, is the subsoil. There are different agricultural characteristics which present distinctive setups for WUSN applications based on the features of the topsoil and subsoil section soils. Water content and soil texture may be different in these soil sections [16], which further affect the wireless communication channel significantly. Further, reflection from the surface of the ground is better in the nodes which are buried at the topsoil section. Mechanical activities like plowing and other such activities take place at the topsoil section of the soil. Therefore, the burial depth is required to be higher than the topsoil section for many agriculture applications, like in crop watering management, and it should be different for applications like border guard. So, for various WUSN applications, sensor nodes are required to be buried at the subsoil section.

WUSNs can be categorized on the basis of positioning section: if the WUSN is installed in the topsoil section, it is called topsoil WUSN and, if it is installed in the subsoil section, it is known as subsoil WUSN as shown in Figure 6.2. In WUSNs, in order to retrieve data, it is required to have the nodes above ground together with the underground buried nodes. Therefore, there are two types of communications that exist in WUSNs as presented in Figure 6.3. Underground-to-underground communication is the interchange of information between the concealed sensor nodes for the purpose of data relay and network management. Underground-to-aboveground communication is then used to collect data in the network. It includes exchanging command and control information from stations which are aboveground to that which are underground. There are different challenges based on these two types of communications. Quality of communication in underground-to-aboveground communication is superior to underground-to-underground communication because some amount of communication takes place above the air [16]. Design and deployment of WUSNs for underground-to-underground communication is generally more challenging.

6.5   LATEST ADVANCES OF WUSNS

In this section, we will focus on the latest advances associated with the underground wireless communication networks. Most of these developments are not basically WUSN applications, but they provide the insight into the challenges related to these networks in the potential areas of applications of Wireless USNs.

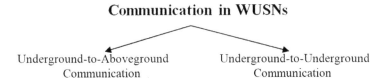

FIGURE 6.3   Classification of communication in WUSNs.

Wireless Underground Sensor Networks are used to observe the underground mines for ensuring security of the mine personnel [17, 18]. As we know that mines are underground, but communication in the sensor nodes takes place through air inside the tunnel or mine. On this basis, features of the wireless channels in the tunnels have been investigated [3, 14].

Landslides can be predicted by using a shallow depth WSN [19]. In this network, MICA2 sensor nodes [12] which are attached with stain gauges and are capable of operating at low depth extent (25–30 cm). In this scheme also, the sensors are concealed under the ground, but the communication takes place in the air. On a similar basis, a sensor network has been constructed to detect the volcano activities. In this application also, for reliable link setup, the antenna of the sensors has to be placed above ground [20].

Another important project which has gained interest in WUSNs is Structural Health Monitoring (SHM). Wisden, a data acquisition system for SHM [21, 22] and Duranode [23] are the two models of WSN application. Other underground systems like Sewers also need structural monitoring but, they require communicating through air for their proper working.

Another usage of WUSNs is the Emscher sewer system which is the biggest domestic water managing project in Europe which uses sensors for gathering necessary information for assessment and cleaning [24]. Communication between the buried sensor nodes and the nodes which are aboveground take place through the phenomenon of radiation [25]. In most of the underground systems, air is used as a communication medium between the sensors.

WUSNs have also been used to deploy a sensor system-based glacier monitoring network, installed in Norway [25]. The main aim of this scheme is to monitor and to measure the constraints of glaciers and ice covers by the use of sensors below the glaciers. In this network, base stations are connected to two wired transmitters and receivers below 30 m of ground level, to avoid wet ice. High transmission power of about 100 mW is used for making the communication between under-ice sensors and the sensors which are places deeper around 80 m from the ground level, possible. This is not a routine underground scenario, but this represents the analogous type of challenges as there are in WUSNs.

Also, there are many experiments being done based on the electromagnetic wave propagation over rocks and soil. It has been considered that in soil, communication takes place by electromagnetic wave transmission using the cable and trials have been done at the frequency of 1–2 GHz, providing a transmission model [26].

Furthermore, tests were performed using ground-piercing radars. This particular study is known as Microwave Remote Sensing (MRS) [27]. By using principles of the ground-penetrating radar, reduction in signal strength and comparative permittivity of numerous constituents, like soil, were determined at a frequency of 100 MHz [8]. MRS is basically used for finding the landmines, by determining the differences among the dielectric constants of soil and the landmine. It was demonstrated that the detection of landmines by Ground Penetrating Radar (GPR) is significantly affected by the soil composition [8]. The above applications give a significant understanding of EM wave propagation through soil, but, no prevailing work gives a widespread representation of concealed communication.

6.6 CHARACTERISTICS OF WUSNS

The characteristics of IoUT-based WUSNs are very much different from other types of wireless sensor networks. This is because the sensor nodes in these networks are buried underground in the soil, because of which these wireless underground channels are influenced by the characteristics of the soil including water content in the soil, its density, temperature, and particle size. As we know that the signals get attenuated primarily because of reflection, absorption, and refraction, when they transmit through multiple environments. WUSNs include three different types of communication channels including Underground to Aboveground Channel (UG-AG), Aboveground to Aboveground Channel (AG-AG), and Underground to Underground Channel (UG-UG), [29, 30]. In soil and air, transmission features of electromagnetic waves are completely dissimilar. There are numerous environmental parameters of soil, like the configuration of soil-related to its constituents, including fractions of clay, sand, and silt, the volumetric water content, the unit concentration and the bulk density. Additionally, attenuation of EM signals in soil obstructs the communication distance which poses a serious challenge to the lifetime of WUSN. Time elapsed between the instant when the network is deployed to the time when it becomes non-functional is called Network lifetime. Communication in soil is influenced by various channel effects as well as path loss in soil. Channel characteristics are also affected by the depth of sensor nodes and multi-path spreading and fading.

Now, the major problems in underground sensor network communication and their possible remedies are discussed below. The sensor depth from the ground surface is generally less than 0.5 m which causes serious problems of signal reflection from the ground. In Figure 6.4, there are two paths of signal propagation: one is direct path and the other is the secondary path because of the reflection of the signal at the ground surface. Out of these, the straight path component represents the main constituent of the received signal, but the reproduced signal also affects communication, particularly when the sensor nodes have low burial depths. Naturally, burial depth of the nodes has a direct impact on the reflection effect. At high burial depths, the reflection effect can be ignored, and the channel can be deliberated as a single path [29], but the attenuation increases which should be considered for most applications. This happens because of the path length of the reflected signal is greater than that of the direct signal, it arrives at the receiver node with less strength due to soil attenuation. In this case, we have to consider only the direct path.

However, if the sensor nodes are concealed near the ground level, the signal reflected by the ground surface needs to be considered, because the received returned waves are strong enough at the receiver, along with the direct waves. This set-up is usually the case of Topsoil. Here, the overall path loss of dual-path channel prototype obtained from [32] is:

$$L_f \text{ (dB)} = L_p \text{ (dB)} - V_{dB} \text{ [32]} \tag{1}$$

where; L_p is the path loss due to the first path in dB; and V_{dB} is the reduction factor of signal due to the second path in dB, i.e.:

$$V_{dB} = 10 \log V \qquad (2)$$

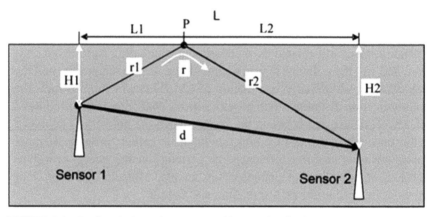

FIGURE 6.4   Dual-path channel prototype with ground reflection.
Source: Adapted from: Ref. [32]. © 2001 Kluwer Academic Publishers.

The dual-path channel prototype defined above signifies the main underground EM wave propagation characteristics. But, in addition to this, there are many other complications in the underground channels which are not depicted in this model. First of all, the ground surface is not actually smooth, which causes reflection and refraction. Secondly, soil has rocks and roots of plants, along with clay which is not consistent. Due to these soil pollutants, multi-path fading can also be deliberated for the underground signal propagation. Many models have been proposed for the underground channel characterization, but because of certain unpredictable nonlinearities, a lot of research is required in this field with main concentration on the weather pattern, geography, flora, and fauna, and other factors which affect the channel.

For above ground conditions, multi-path fading meant for above-air wireless communication has been studied widely [32]. The existence of random effects like fixed and moving obstacles cause variation and bending of the EM signals in air, whereby, amplitude and phase of the received wave shows arbitrary time-varying nature. Concealed channel between two transceivers, does not present any random refractions and is relatively stable. The variation occurs due to the roots of trees, clay particles and rocks, etc., which causes reflection, refraction, and uneven EM wave signals causing underground channels to exhibit similar multipath characteristics as that of air, although causes are different.

6.7 WUSN NODE ARCHITECTURE

Main components of IoUT enabled Wireless Underground Sensor Networks include sensing model which consists of moisture sensor and signal converters like analog-to-digital (A/D) and digital-to-analog (D/A) converters, processors having Central Processing Units (CPU) and storage devices, Communication Model, and a power supply module. With the advancement in technology, these components are becoming more and more economical and smaller. Sensor node must be very reliable and should be able to run unattended for a long time, even for years. This poses stringent power requirements on the network. For the design of power sources of a WUSN, there are many factors that need to be considered: (a) choice of battery type, and (b) choice of small power electronic design schemes [33]. Modularizing design method is used for designing wireless sensor nodes. Architecture of WUSN using this method is shown in Figure 6.5. All the nodes are composed using this system structure which comprises the sensor unit, processor unit, energy supply unit and wireless communication unit.

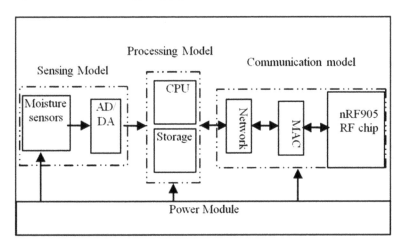

FIGURE 6.5   Wireless sensor network node architecture.

Source: Reprinted from Ref. [35]. © 2012 African Journal of Biotechnology.
https://creativecommons.org/licenses/by/4.0/

The design consists of digital sensor DS18B20 which collects information about the soil moisture content. DS18B20 is a digital temperature sensor, produced by the United States Company Dallas. This microprocessor consumes low power, small in size, has high performance and has strong ability for anti-interference and can match easily [34]. 16-bit MSP430 is used by microcontroller, which is very economical, has a high level of integration and is used in low power applications. Wireless chip nRF905 is also used in this architecture to collect and transmit complete information in WUSN. Architecture of the Sink node is same as that of WUSN node. Here, only sink node is above the ground, rest all the nodes are underground. All underground nodes communicate with the sink node by transmitting information to the ground sink node, which covers the network better [35].

Node set of WUSN depends on the particular application. It can be set as different layer, presented in the equal or different depths. To keep the sink node in the range of communication, it is kept static or mobile. The topology structure is shown in Figure 6.6.

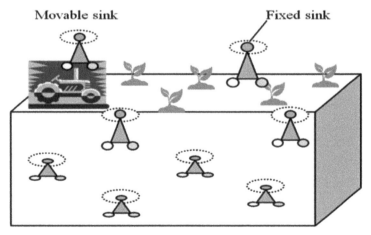

FIGURE 6.6   Topology structure of wireless sensor network.
Source: Adapted from: Ref. [31]. © 2007 Elsevier.

6.8   COMMUNICATION ARCHITECTURE

Figure 6.7 shows the WUSNs protocol stack. It demonstrates the classical five-layer protocol hierarchy, along with the two cross-layered planes for power and task management.

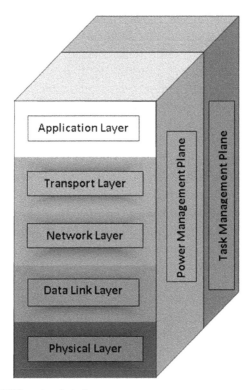

FIGURE 6.7 WUSN protocol stack.

The terrestrial WSN protocols cannot address the exceptional demands of the concealed environment. Therefore, to guarantee the efficient and reliable functioning of WUSNs, it is essential to reconsider and rework on each of the layers. In addition, cross-layered design enhances the efficiency of the protocol stack. Before discussing the cross-layered design of WUSNs, we will examine the research challenges to be addressed on each level of the WUSN protocol stack to make it feasible [2].

6.8.1 PHYSICAL LAYER

To make communication at the physical layer of WUSNs possible is a very challenging task. Losses are very high when EM wave signals propagate via rock and soil. The underground environment is dynamic in nature which is another big challenge. Additionally, loss rates have high dependency on various soil attributes, specifically moisture content [8], which may fluctuate with time. Wet soil highly attenuates the EM waves, which even makes communication impossible at any distance [26]. For example, after a rainfall, there is a rise in soil moisture level which produces losses that can last for a significant amount of time.

It is very important to design an efficient antenna for propagation of EM waves in the underground, because of the challenging nature of underground communication. The radiation and reception characteristics can be significantly affected by implanting an antenna in soil which is a conductive medium [2]. It has been seen that when EM waves pass through the soil and rock, their attenuation is directly proportional to their frequency [2]. So, for making communication feasible at several meters, lower frequencies should be used. Traditional EM antennas cannot be used at the low frequencies because of their larger size. The selection of the most suitable modulation technique in WUSNs is another challenge, for the given power constraints and low frequencies used to decrease path losses. Less bandwidth available because of the use of lower carrier frequencies means lower data rates than terrestrial WSNs. Data rate in WUSNs are further reduced by the extreme channel losses.

There are many research challenges that need to be addressed at physical layers. Firstly, it is required to find the most promising physical layer scheme by further analyzing the electromagnetic induction and vibrational signaling in the underground. Then, finding the power-efficient modulation schemes for underground communication is the area of research in WUSNs. The agreement between capacity and reliability must be considered, because, losses are decreased by lower frequencies, but it also provides lesser bandwidth which reduces the channel capacity. Moreover, the capacity of the underground wireless communication channels needs to be studied theoretically.

6.8.2　DATA LINK LAYER

MAC layer protocols [31] designed for land-dwelling WSNs are not going to work in WUSNs, because they are either TDMA based or connection-oriented and concentrate on decreasing the consumption of energy by focusing on four main fundamental areas like overhearing, collisions, idle listening and controlling overhead of packet [23]. Because of the different features of wireless underground channels, extraordinary MAC protocols are required. In existing MAC protocols for terrestrial WNSs, the main concern is to conserve energy, which is accomplished by reduction of idle listening time [31]. But, due to high path losses incurred in WUSNs in the ground, transmission power required is much higher than in terrestrial WSNs. For acceptable lifetime of underground sensor networks, number of transmissions should be less. But, because of collisions, signal need to be retransmitted. Thus, WUSN MAC protocols should be designed to avoid collisions. It is possible by designing contention-based protocols which use RTS/CTS scheme, but that again has the drawback of introducing unacceptable overhead in WUSNs. TDMA-based scheme reserves a timeslot for each device to transmit, and thus eliminates collisions. But it again introduces large overhead and also produces synchronization problem. Sensor data needs to be reported infrequently in WUSN devices, so, power can be saved by operating them at low duty cycle. But, due to the large drift in the device's clock during sleep periods, synchronization in the network may lose. Due to energy constraints of sensor devices and high losses incurred in the underground channels, received signals may have moderately high bit error rates (BER). Moreover, ARQ technique cannot be used for WUSNS, because of the large overhead of acknowledgments and energy consumption during packet retransmission. It will be better to choose FEC channel coding schemes, but still, it requires research in this field.

The main research challenges at data link layer are to find whether a contention-based or a TDMA based MAC protocol is more suitable for Wireless USNs, designing an optimized hybrid MAC structure most suitable for the underground networks, to explore the achievement of synchronization, to explore powerful Forward Error Correction methods for underground channels and to maximize power efficiency, to find optimal packet size and to provide good quality of service with power conservation [2].

6.8.3 NETWORK LAYER

Routing protocols for ad-hoc networks are generally categorized as: reactive, proactive, and geographical. Reactive routing protocols work only when the route is required to be discovered, while proactive protocols work by continuously maintaining routes among every device in the link. Both these protocols result in high overhead. Moreover, because of high water content in soil, it is highly likely that the device may be unreachable or the preset routes like in proactive routing may not always be beneficial because the network may not be synchronized overextended sleep times. On the other hand, in geographical routing protocols, routes are established by interpreting the information about the actual position of the devices. Therefore, the route created with each hop makes information substantially nearer to the endpoint. Deployment of geographical routing protocols determines its usefulness. To deploy most of the sensors, holes are drilled in the soil and at the time of deployment, detailed information about the location of each hole is recorded. In this way, geographical routing protocols may be beneficial. Another method of WUSN deployment is to cover the randomly scattered sensing devices with soil. But, in this way, the exact location of the sensor devices is not known which can suppress the usefulness of geographical protocols [2].

In terrestrial WSNs current routing protocols, all the devices can equally participate in determining the path between the source and the sink. Most of them deliberate the existing energy level of the node and the radio transmission power required for communication among any two devices in a WUSN can fluctuate significantly. If the data is routed through sensors in a WUSN deployed in open-air mine, it will be more efficient than the devices which are implanted in the soil and rock. Also, the cost of link varies with variation in water content of soil. To maximize the lifetime and power efficiency, challenges of underground networks should be known.

Network layer protocols must address the issues like efficient handling of the changes in the network topology made by the abrupt changes in the sensing intervals, effects on routing protocols due to low duty cycle of WUSNs, monitoring underground pressure sensors for security issues in time-sensitive underground sensor networks and to make multi-path routing algorithms more energy-efficient [2].

6.8.4   *TRANSPORT LAYER*

The flow and error control features are incorporated in WUSNs through a transport layer protocol. Again, because of the high loss rate of underground channels, the transport layer protocols designed for terrestrial WSNs need to be re-examined. The main focus should be on increasing the network efficiency and to save the limited sensor resources. Reliability is another important feature of these protocols to guarantee the correct identification of the attributes of the event as expected by the sensor networks. Flow control is required to prevent network devices from flooding the network with data transmission in excess of the network capacity and memory. Network should be prevented from congestion by transmitting excessive data. But, congestion near the sink is a common problem in WUSNs because of their low data rates. This problem can be solved by routing the data to terrestrial relays which have large data rate and this is possible at the network layer. This is called cross-layering solution to this jamming problem. Prevailing TCP protocols incorporate flow control based on either retransmission or window-based protocols, which is unsuitable for WUSNs. But, in WUSNs, number of retransmissions should be reduced to save energy. Due to highly unreliable underground channels, completely new strategies should be devised for reliability and flow control [2].

Research challenges at this layer include the development of new mechanisms to decrease the packet losses, based on the underground network event model, optimal policies for congestion control and increasing throughput efficiency for transport reliability in limited bandwidth networks, needed to define acceptable loss rate in WUSNs, and research is also needed at transport layer to provide varied levels of service for various sensor data types.

6.8.5 CROSS-LAYERING

Cross-layered protocol design has to address various design challenges as given below:

1. **Channel Prediction by Using Sensor Data:** As we know from the above discussion that the soil water content of underground networks varies at times, and it has a direct impact on the condition of the underground channel. To monitor this variability, moisture sensors are used in a large number of sensor devices in WUSN network. This gives the concept of cross-layering within the application layer where the readings of moisture content are taken directly, and the radio output power of lower layers is adjusted accordingly, to choose routes correctly, and to choose the most efficient FEC scheme.

2. **Predicting Soil Property by Using Channel Data:** Channel properties can also be predicted by the sensor readings. In this method, losses are increased slowly in the channel between two

devices to interpret the increase in water content in the soil. Other devices remain reachable. This can be used to predict the condition of soil between the devices where there are no sensors deployed and it provides cross-layer interaction between the network and application layers.

3. **Routing based on Physical Layer:** Cross-layering, MAC, and routing solution can result in power savings. In WUSNs, it is necessary to communicate with the device neighbors at different power levels due to wide variations of soil conditions over short distances. For increasing the network lifetime, low transmission power links need to be utilized. Physical layer gathers this information and passes it to the network layer. Additionally, packets are routed through dry areas where there is less attenuation by soil, by taking readings from surrounding devices and then processing them to form the water content map of the deployment terrain of the network [2].

4. **Resourceful Scheduling at the MAC Layer:** Sensor data at the application layer is useful in accomplishing this. Suppose, soil moisture content is increasing continuously in the sensor reading, the device will try to reduce the further losses by routing the packets at higher power level, and inserts period of silence in between outbound packets and waits for the drop in soil water content, so that, power can be conserved. After the water content is decreased, it will need lower transmission power and a fewer retransmissions.

5. **Cross-Layering between Data Link and Transport Layers:** In a cross-layering unified unit, functions of the transport layer can be strongly unified with that of the data link layer. This integration will make the information about the varying concealed channel condition accessible to the transport layer. Because of this cross layering, the channel information is made transparent to the higher layers, which was otherwise not available. This maximizes the transport proficiency, and can help transport and data link layer behavior to adjust dynamically according to the varying underground behavior [2].

6.9   APPLICATIONS

The potential underground network applications can be classified into four main categories: infrastructure monitoring, environment monitoring, location determination, and border patrolling and security monitoring.

6.9.1 INFRASTRUCTURE MONITORING

WUSNs can be used to monitor a large amount of underground infrastructure such as electrics, pipes, and liquid storing reservoirs. Consider a fuel station which stores fuel in underground reservoirs that needs to be cautiously supervised, so that there is no leakage in the underground tanks, and continuous monitoring is also required to find the volume of fuel in the reservoir.

Sometimes, the sewer system is not there in some locations. At these locations, homes have an underground putrefying chamber, which needs to be monitored by WUSNs to avoid overflow.

For monitoring underground drainage system, sensors can be installed alongside the conduit of pipelines to quickly locate and repair the leakages.

WUSNs can also be used to monitor the operational health of any buried components of a dam, building, or bridge [23]. Wireless sensor devices can be implanted within vital structural constituents to observe strain, stress, and other factors [28].

In addition, WUSNs can be advantageous for military uses like in minefield monitoring where an underground infrastructure exists.

6.9.2 ENVIRONMENTAL MONITORING

As per the above discussion, sensor nodes are useful in agriculture to observe the factors like moisture and mineral amount in the underground soil, and to make available the exact data for watering and fertilization. WSN can offer extensive help in soil care like it can provide information regarding which particular area of field requires moisture instead of irrigating the whole field. Sensors can also be deployed in each pot in a greenhouse setting.

Compared to current terrestrial WSNs, WUSNs offer concealment, which makes it more attractive for monitoring sports field, to monitor condition of soil at soccer fields, golf courses (see Figure 6.8), grass tennis courts and baseball fields. In all these applications, soil conservation is very essential to keep grass in good physical shape, because of poor lawn conditions. Underground sensors are also protected from tractors and lawn-movers.

Another possible application of WUSNs is for monitoring the existence of several deadly materials in soil near river beds and artesian basins, where the presence of such chemicals could infect drinking water supplies.

In such circumstances, a hybrid network consisting of underground and underwater sensors is utilized.

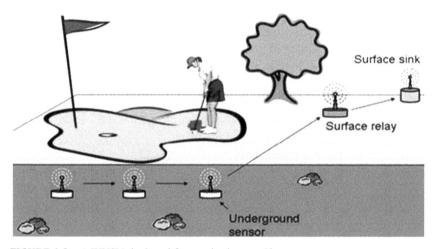

FIGURE 6.8   A WUSN deployed for monitoring a golf course.
Source: Reprinted with permission from Ref [2]. © 2006 Elsevier.

WUSNs are also useful in monitoring soil movement to predict possible landslide [19]. For improved landslide prediction, WUSN technology allows for a deeper placement of sensors, so that it could be possible to warn the residents of affected areas well in time, to evacuate.

Air quality in underground coal mines can also be monitored using WUSNs. Thus, explosions in the mines can be avoided by continuously monitoring the accumulation of carbon monoxide and methane in the mine [15]. Another mining application of underground sensor nodes is to locate and rescue trapped miners.

Prediction and monitoring earthquakes can also be assisted by WUSN technique. Useful information for earthquakes is gathered from many depths further down the ground. Here, the multiple-hop property of WUSNs is used.

6.9.3   *LOCATION DETERMINATION OF OBJECTS*

For location-based services, static underground sensors that know their location can be used as a beacon. For example, consider a network in which underground sensor devices below the road surface can converse with a car driving over it. It can alert the driver for the traffic signal or upcoming stop signal. This information could be received by the car and conveyed to the driver.

To provide navigational support for self-directed systems, like in a self-sufficient fertilizer unit, to navigate in the region to be pollinated based on underground sensors, location determination can be utilized.

Another use of WUSN technology is in the location of people in a collapsed building. Sensors could be deployed and programmed within the building. Persons of the building could carry the programmed device. During collapse, the individual's device could be located by communicating with the static devices to rescue the survivors [2].

6.9.4 SECURITY MONITORING AND BORDER PATROLLING

WUSNs may be useful in monitoring the presence and movement of objects or people above the ground. Like in location determination, sensors must be immobile and should be conscious of their position. But, distinct from location finding, different types of magnetic, pressure or acoustic sensors are required to govern the presence of an individual or an object. This could be useful for intrusion detection at business and home security. Underground sensor nodes could be installed everywhere around the structure to detect impostors.

WUSNs can be used for border patrolling by deploying wireless pressure sensors along the border side to make the authorities aware of unlawful passages [2].

6.10 CONCLUSION

In this chapter, the idea of Wireless Underground Sensor Networks has been discussed. Sensor devices are arranged completely underneath the surface in WUSNs and their communication is facilitated by IoUT. They have a wide range of applications in military, underground sensing, testing soil traits and moisture content, pollution control, location detection, security, and detection of natural calamities. WUSNs have vast scope of improvement and development without major changes in the existing data collection and monitoring approach. In spite of their huge prospective, the growth of WUSNs has to cope with several research challenges due to the complex concealed environment. Underground environment is mainly a tough environment for wireless transmission because the underground channel condition depends on the features like moisture in the soil or rock in which nodes are installed. It was also discussed that low frequencies transmit with lesser attenuation through the concealed networks, but they severely restrict the bandwidth offered for data communication in WUSNs. Low data rates together with high losses in underground channels describe the significance of sustaining energy and demand the re-examination of prevailing earthly WSN communication standards for developing new standards for WUSN networks. Key characteristics and classification of underground sensor networks has been reviewed. Potential application areas have been discussed. Most important research opportunities at every layer of the protocol hierarchy have also been presented with the scope of improvement at each layer of WUSN. This study can lay the foundations of improvement in subversive communication and will also help in the development of newer applications and future research of wireless underground sensor networks. In future, the sensor networks will be a vital part of our lives.

KEYWORDS

- aboveground
- bit error rate
- electromagnetic
- ground penetrating radar
- Internet of Things
- internet of underground things
- structural health monitoring
- underground
- wireless sensor networks
- wireless underground sensor networks

REFERENCES

1. Younis, M., (2018). Internet of everything and everybody: Architecture and Service Virtualization. *Comput. Commun., 131,* 66–72.
2. Akyildiz, I. F., & Stuntebeck, E. P., (2006). Wireless underground sensor networks: Research challenge. *Ad. Hoc. Networks Journal, 4,* 669–686. Elsevier.
3. Akyildiz, I. F., Sun, Z., & Vuran, M. C., (2009). Signal propagation techniques for wireless underground communication networks. *Physical Communication Journal, 2,* 167–183. Elsevier.
4. Saeed, N., Alouini, M. S., & Al-Naffouri, T. Y., (2019). Toward the internet of underground things: A systematic survey. *IEEE Commun. Surv. Tutor., 21,* 3443–3466.
5. Vuran, M. C., & Silva, A. R., (2010). Communication through soil in wireless underground sensor networks-theory and Practice. *Sensor Networks,* 309–347.
6. Ali, H., & Choi, J., (2019). A review of underground pipeline leakage and sinkhole monitoring methods based on wireless sensor networking. *Sustainability, 11,* 1–24.
7. Imanishi, K., Ellsworth, W., & Prejean, S. G., (2004). Earthquake source parameters determined by the SAFOD pilot hole seismic array. *Geophysical Research Letters, 36.*
8. Bala, I., Bhamrah, M. S., & Singh, G., (2017). Rate and power optimization under received-power constraints for opportunistic spectrum-sharing communication. *Wireless Personal Communications, 96,* 5667–5685. Springer.
9. Bala, I., Bhamrah, M. S., & Singh, G., (2017). Capacity in fading environment based on soft sensing information under spectrum sharing constraints. *Wireless Networks, 23,* 519–531. Springer.
10. Cardell-Oliver, R., Smettem, K., Kranz, M., & Mayer, K., (2005). A reactive soil moisture sensor network: Design and field evaluation. *International Journal of Distributed Sensor Networks, 1,* 149–162.
11. Chehri, A., Fortier, P., & Tardif, P. M., (2006). Application of ad-hoc sensor networks for localization in underground mines. *IEEE Annual Wireless and Microwave Technology Conference,* 1–4.
12. Behari, J., (2005). *Microwave Dielectric Behavior of Wet Soils.* Springer.
13. Mastarone, J. F., & Chappell, W. J., (2006). Urban sensor networking using thick slots in manhole covers. *IEEE Antennas and Propagation Society International Symposium,* 779–782.
14. Sun, Z., & Akyildiz, I. F., (2008). Channel modeling of wireless networks in tunnels. *IEEE Globecom '08,* 1–5.
15. Dubaniewicz, T., Chilton, J., & Dobroski, H., (1991). Fiber optics for atmospheric mine monitoring. *IEEE Industry Applications Society Annual Meeting* (Vol. 2, pp. 1243–1249). *Dearborn, MI.*
16. Stuntebeck, E., Pompili, D., & Melodia, T., (2006). *Wireless Underground Sensor Networks using Commodity Terrestrial Motes* (pp. 112–114). SECON.
17. Chehri, A., Fortier, P., & Tardif, P., (2007). Security monitoring using wireless sensor networks. *Fifth Annual Conference on Communication Networks and Services Research (CNSR '07),* 13–17.
18. Kennedy, G. A., & Bedford, M. D., (2014). Underground wireless networking: A performance evaluation of communication standards for tunnelling and mining. *Tunn. Undergr. Space Technol., 43,* 157–170.

19. Sheth, A., et al., (2005). Senslide: A sensor network-based landslide prediction system. *SenSys'05: 3rd International Conference on Embedded Networked Sensor Systems,* 280–286.
20. Werner-Allen, G., et al., (2006). Deploying a wireless sensor network on an active volcano. *IEEE Internet Computing, 10,* 18–25.
21. Paek, J., et al., (2005). A wireless sensor network for structural health monitoring: Performance and experience. *IEEE Conference on Embedded Networked Sensors,* 1–9.
22. Xu, N., et al., (2004). A wireless sensor network for structural monitoring. *ACM Conference on Embedded Networked Sensor Systems* (pp. 13–24). Baltimore, MD.
23. Park, C., Xie, Q., Chou, P., & Shinozuka, M., (2005). Duranode: Wireless networked sensor for structural health monitoring. *IEEE Sensors,* 277–280.
24. Elkmann, N., et al., (2007). Development of fully automatic inspection systems for large underground concrete pipes partially filled with wastewater. *Robotics and Automation,* 130–135.
25. Martinez, K., Ong, R., & Hart, J., (2004). Glacsweb: A sensor network for hostile environments. *IEEE Communications Society Conference on Sensor and Ad Hoc Communications and Networks, SECON '04,* 81–87.
26. Weldon, T. P., & Rathore, A. Y., (1999). *Wave Propagation Model and Simulations for Landmine Detection* (pp. 1–17). Tech. Rep., University of North Carolina at Charlotte.
27. Daniels, D. J., (1996). Surface-penetrating radar. *Electronics & Communication Engineering Journal, 8,* 165–182.
28. Cheekiralla, S., (2004). *Development of a Wireless Sensor Unit for Tunnel Monitoring.* Master's thesis, Massachusetts Institute of Technology.
29. Bala, I., Bhamrah, M. S., & Singh, G., (2019). Investigation on outage capacity of spectrum sharing system using CSI and SSI under received power constraints. *Wireless Networks, 25,* 1047–1056. Springer.
30. Akkas, M. A., Akyildiz, I. F., & Sokullu, R., (2012). Terahertz channel modeling of underground sensor networks in oil reservoirs. *IEEE Ad. Hoc. and Sensor Networking Symposium,* 543–548.
31. Kredo, K., & Mohapatra, P., (2007). Medium access control in wireless sensor networks. *Computer Networks, 51,* 961–994. Elsevier.
32. Stuber, G., (2006). *Principle of Mobile Communication.* Kluwer Academic Publishers, 2/e.
33. https://www.cambridge.org/core/terms. https://doi.org/10.1017/CBO9781139030960. 002 (accessed on 16 November 2021).
34. Fezari, M., & Dahoud, A. A., (2019). *Exploring One-Wire Temperature Sensor "DS18B20" with Microcontrollers.*
35. Yu, X., Wu, P., Han, W., & Zhang, Z., (2012). Overview of Wireless Underground Sensor Networks for Agriculture. *African Journal of Biotechnology, 11,* 3942–3948.

CHAPTER 7

Computational Intelligence-Based Energy Efficient Clustering Protocol for Underground Wireless Sensor Networks

PALVINDER SINGH MANN[1] and SATVIR SINGH[2]

[1]DAV Institute of Engineering and Technology, Jalandhar, Punjab–144008, India, E-mail psmaan@davietjal.org

[2]I.K. Gujral Punjab Technical University, Punjab–152004, India

7.1 INTRODUCTION

Over the last few years, underground wireless sensors networks (UWSNs) witnessed tremendous growth due to low-cost, multi-functional capabilities even in diverse and complex environments such as habitat monitoring, deep-sea remote sensing, to name a few, which is part of the internet of underground things (IoUT). UWSNs contain self-configured, distributed, and autonomous sensor nodes (SNs) that monitor physical or environmental activities like humidity, temperature, or sound in a specific area of deployment [37]. SNs can have more than one sensor to capture data from the physical environment, wherever deployed. A sensor with limited storage and computation capabilities receives the sensed data through an analog to digital converter (ADC) and process it further for transmission to the main location, known as *Sink* or *Base Station* (BS), where the data can be analyzed for decision making in variety of applications [2]. Every node also acts as a repeater for passing information of other sensor

Harnessing the Internet of Things (IoT) for Hyper-Connected Smart World.
Indu Bala and Kiran Ahuja (Eds.)

nodes to the sink. The most important part of the sensor node is its power supply, which caters to the energy requirements of sensors, processors, and transceiver, however, its limited battery life can lead to premature exhaust of the network due to excessive usage [1]. As manual recharging of batteries is not possible in complex deployments, efficient use of the energy becomes a tough challenge in applications where a prolonged life of the network is required [9]. Researchers are heavily involved in designing of energy-efficient solutions, however, on the other hand, network life can also be extended by planning energy-efficient approaches. It is well accepted that cluster based hierarchical approach is an efficient way to save energy for distributed WSNs [32], which increase network life by effectively utilizing the node energy and support dynamic WSNs environment. In a cluster based WSNs, SNs are divided into several groups known as clusters with a group leader known as *Cluster Head* (CH). All the SNs sense data and send it to their corresponding CHs, which finally send it to the BS for further processing. Clustering has various significant advantages over conventional schemes. First, data aggregation is applied on data received from various SNs within a cluster, to reduce the amount of data to be transmitted to BS; thus, energy requirements decrease sharply. Secondly, rotation of CHs helps to ensure a balanced energy consumption within the network, which prevents getting specific nodes starved due to lack of energy [5]. However, selection of appropriate CH with optimal capabilities while balancing the energy-efficiency ratio of the network is a well-defined multi-modal optimization problem in WSNs [16]. Thus, metaheuristic-based approaches including Evolutionary algorithms (EAs), Particle swarm optimization (PSO), Genetic algorithm (GA), and recently, Artificial bee colony (ABC), have been used extensively as population-based optimization methods by different researchers for energy-efficient clustering protocols in WSNs [28]. Results shows that the performance of the ABC metaheuristic is competitive to other population-based algorithms with the advantage of employing fewer control parameters with simplicity of use and ease of implementation [18]. However, similar to other population-based algorithms, the standard ABC metaheuristic also face some challenges, as it is considered to have poor exploitation cycle than exploration, moreover convergence rate, is typically slower, specially while handling multi-modal optimization problems [13]. Therefore, an improved Artificial bee colony (iABC) metaheuristic is presented, with an improved solution search equation, which will be able to search an optimal solution to improve its exploitation capabilities and an improved approach for population sampling through the use of first of its kind compact Student's *t* distribution to enhance the global convergence of the proposed metaheuristic. Further, to utilize the capabilities of the proposed metaheuristic, an improved artificial bee colony based clustering protocol (*iABC*2) is introduced, which selects optimal cluster heads (CHs) with an energy-efficient approach for WSNs.

7.2   RELATED WORK

A large number of clustering protocols have been developed so far for UWSNs. Here we present only the vital contributions of the researchers based on conventional as well as metaheuristic approach; however, we underline the significance of metaheuristic, as our proposed protocol is part of this approach. Low-energy adaptive clustering hierarchy (LEACH) [10], is a conventional clustering protocol which combines energy-efficient cluster-based routing to application-oriented data aggregation and achieve better lifetime for a WSN. LEACH introduces an algorithm for adapting clusters and rotating CHs positions to evenly distribute the energy load among all the SNs, thus enabling self-organization in WSNs. LEACH remain a paradigm architect for designing clustering protocols for WSNs till date. HEED (Hybrid Energy-Efficient Distributed clustering) [38], is another conventional clustering protocol that selects CHs based on hybridization of node residual energy and node proximity to its neighbors or node degree thus achieving uniform CH distribution across the network. HEED approach can be useful to design WSNs protocols that require scalability, prolonged network lifetime, fault tolerance, and load balancing, but it only provided algorithms for building a two-level hierarchy, and no idea is presented for designing protocol to multilevel hierarchies. Power-efficient and adaptive clustering hierarchy (PEACH) [36], selects CHs without additional overhead of wireless communication and supports adaptive multi-level clustering for both location-unaware and location-aware WSNs but with high latency and low scalability, thus make it suitable only for small networks. T-ANT [29], a swarm-inspired clustering protocol which exploit two swarm principles, namely separation and alignment, through pheromone control to obtain a stable and near-uniform distribution for selection of CHs. Energy-Efficient

Multi-level Clustering (EEMC) [12], achieve less energy consumption and minimum latency in WSNs by forming multi-level clustering with minimum algorithm overhead. However, the authors ignored the issue of channel collision which happens frequently in wireless networks. Energy-efficient heterogeneous clustered scheme (EEHC) [17], selects CHs based on weighted election probabilities of each node, which is a function of their residual energy and further support node heterogeneity in WSNs. Multi-path Routing Protocol (MRP) [35], is based on dynamic clustering with Ant colony optimization (ACO) metaheuristic. A CH is selected based on residual energy of nodes, and an improved ACO algorithm is applied to search multiple paths that exist between the CH and BS. MRP prolonged the network lifetime and reduces the average energy consumption effectively using proposed metaheuristic. Energy-Efficient Cluster Formation protocol (EECF) [5], presents a distributed clustering algorithm where CHs are selected based on a three-way message exchange between each sensor and its neighbors while possessing maximum residual energy and degree. However, the protocol does not support multi-level clustering and consider small transmission ranges. Mobility-based clustering (MBC) protocol [7] support node mobility, hence CHs will be selected based on nodes residual energy and mobility, whereas a non-CH node maintains link stability with its CH during the set-up phase. UCFIA [20], is a novel energy efficient unequal clustering algorithm for large scale WSNs, which use fuzzy logic to determine node's chance to become CH based on local information such as residual energy, distance to BS and local density of nodes. In addition, an adaptive max-min ACO metaheuristic is used to construct energy-aware inter-cluster routing between CHs and BS, thus balances the energy consumption of CHs. Distributed Energy-Efficient Clustering with Improved Coverage (DEECIC) [19], selects the minimum number of CHs to cover the whole network based on nodes local information and periodically updates CHs according to nodes residual energy and distribution. By reducing overheads of time synchronization and geographic location information, it prolongs network lifetime and improve network coverage. Energy-Aware Evolutionary Routing Protocol (ERP) [3], is based on evolutionary algorithms (EAs) and ensures better trade-off between lifetime and node stability period of a network with efficient energy utilization in complex WSNs environment. Harmony search algorithm based clustering protocol (HSACP) [11], is a centralized clustering protocol based on harmony search algorithm (HSA), a music-inspired metaheuristic, which is designed and implemented in real-time for WSNs. It is designed to minimize the intra-cluster distances between the cluster members and their CHs thus optimize the energy distribution for WSNs. Kuila and Jana [15] presents a linear/nonlinear programming (LP/NLP) formulations of energy-efficient clustering and routing problems in WSNs, followed by two algorithms for the same based on a particle swarm optimization (PSO). Their proposed algorithms demonstrate their proficiency in terms of network life, energy consumption, and delivery of data packets to the BS.

Conventional approaches [4] are better in self-organization, load balancing with minimum overhead but average in energy-efficiency whereas metaheuristic shown to be good in energy-efficiency with prolong network life. Therefore, metaheuristic-based approaches need to be further explored and improved for energy-efficiency solutions in WSNs.

7.3   ARTIFICIAL BEE COLONY (ABC) METAHEURISTIC

Original Artificial bee colony (ABC) metaheuristic is proposed by Karaboga [13] for optimizing multi-variable and multi-modal continuous functions, which has aroused much interest in the research community due to less computational complexity and use of less number of control parameters. Moreover, optimization performance of ABC is competitive to well-known state-of-the-art meta-heuristics [14]. In ABC, there are three types of bees: employed bees, onlookers, and scout bees. The employed bee carries *exploitation* of a food source and share information like direction and richness of food source with the onlooker bee, through a *waggle* dance, thereafter onlooker bee will select a food source based on a probability function related to the richness of that food source, whereas scout bee *explores* new food sources randomly around the hive. When a scout or onlooker bee finds a new food source, they become employed again, on the other hand, when a food source has been fully exploited, all the employed bees will abandon the site and may become scouts again. In ABC metaheuristic, a food source corresponds to a possible solution to the optimization problem and the number of employed bees is equal to the number of food sources. The detail procedure of ABC metaheuristic in different phases are presented in subsections [39].

7.3.1   INITIALIZATION PHASE

ABC metaheuristic starts with initial population number (PN), randomly-generated through D-dimensional real set of vectors. Let $x_{ij} = \{x_{i1}, x_{i2}, \ldots x_{iD}\}$ is the i-th food source, where $j = 1, 2\ D$, is obtained by:

$$x_{ij} = x_{minj} + rand(0, 1)(x_{maxj} - x_{minj}) \tag{1}$$

where; x_{minj} and x_{maxj} denotes lower and upper limits, respectively.

7.3.2   EMPLOYED BEE PHASE

In this phase, each employed bee obtain a new solution v_{ij} from x_{ij} using expression:

$$v_{ij} = x_{ij} + \varphi_{ij}(x_{ij} - x_{kj}) \tag{2}$$

where; k is randomly obtained from $\{1, 2\ldots, SN\}$; and φ_{ij} is a uniform random number between $[-1, 1]$. The value of v_{ij} is obtained and compared to x_{ij}, further if the fitness of v_{ij} comes out better than x_{ij}, then bee will forget the old solution and remember the new one. Otherwise, it will keep exploiting x_{ij}.

7.3.3   ONLOOKER BEE PHASE

All employed bees share the nectar information of their food source with the onlookers, through a *waggle* dance performed at their hive, after which they select a food source depending on a probability p_i as:

$$p_i = \frac{f_i}{\sum_{n=1}^{SN} f_i} \tag{3}$$

where; f_i is the fitness of x_{ij}. Onlooker bee choose a food sources with higher fitness and search x_{ij} according to Eqn. (2), now if the new solution has a better fitness, it will replace x_{ij}.

7.3.4   SCOUT BEE PHASE

After a number of trials, called maximum cycle number (MCN), if a solution cannot be improved further, then food source is abandoned, and the corresponding employed bee becomes a scout again. The scout will then produce a new food source randomly by using Eqn. (1) again.

7.4   IMPROVED ARTIFICIAL BEE COLONY (*IABC*) METAHEURISTIC

Like standard ABC metaheuristic, its variants too face some challenges, like the convergence rate is typically slow since they find difficulty in choosing the most promising search solution, while solving complex multi-modal optimization problems. To overcome these limitations, we present an improved Artificial bee colony (*iABC*) metaheuristic with an improved initialization phase for better sampling and improved solution search equation, named ABC/rand-to-opt/1 with optimal search abilities. The details of the proposed metaheuristic are as follows.

7.4.1   IMPROVED INITIALIZATION PHASE

Population initialization is an important step in evolutionary algorithms as it can affect the convergence rate and quality of the final solution. Moreover, a large amount of the memory is needed either to store the trial solutions or control parameters of the problem. To reduce the memory requirements, the concept of virtual population has been introduced [8] through the family of Estimation of Distribution Algorithms (EDA) [21] framework by considering compact probability density functions (cPDFs). Therefore, we propose *Student's–t* [22] distribution [34]; a cPDF, which needs only one vector to be stored on memory thus reduces storage and step-up convergence rate. The proposed distribution [23] can be described by Eqn. (4) where $f(x_{ij})$ is the value of the cPDF corresponding to variable x_{ij}, the (∞, ∞) domain of the proposed cPDF is truncated to $[-1, 1]$ and B represents a Beta function. By applying this cPDF [24, 25], only vector κ needed to be stored on memory. This cPDF is being introduced first time in population-based metaheuristic due to its compact nature.

$$f\left(x_{ij}\right) = \frac{\left(1 + \dfrac{x_{ij}^2}{k}\right)^{-(k+1)/2}}{\sqrt{kB}\left(\dfrac{1}{2}, \dfrac{k}{2}\right)} \tag{4}$$

Further, we suggested a new alternative with Cumulative distribution function (CDF) of the proposed *Student's t* distribution, where a pair of cPDFs that share the same parameters are derived through *Student's t* CDF by taking integral from x to x with respect to dx for function $f(x)$ as mentioned below:

$$\int_{-x}^{x} f(x)dx = \frac{1}{\sqrt{\kappa}B\left(\frac{1}{2},\frac{\kappa}{2}\right)} \int_{-x}^{x}\left(1+\frac{x^2}{x}\right)^{-(\kappa+1)/2} dx$$

$$= \frac{2}{\sqrt{\kappa}B\left(\frac{1}{2},\frac{\kappa}{2}\right)} \int_{0}^{x}\left(1+\frac{x^2}{x}\right)^{-(\kappa+1)/2} dx$$

$$= \frac{-2}{\sqrt{\kappa}B\left(\frac{1}{2},\frac{\kappa}{2}\right)} \int_{1}^{\kappa/\kappa+x^2} \frac{x^{\kappa+1/2}\kappa\sqrt{x}}{2x^2\sqrt{\kappa}\sqrt{(1-x)}} dx$$

$$= \frac{1}{B\left(\frac{1}{2},\frac{\kappa}{2}\right)} \int_{\kappa/\kappa+x^2}^{1} (1-x)^{-1/2} x^{\kappa/2-1} dx$$

$$= \frac{1}{B\left(\frac{1}{2},\frac{\kappa}{2}\right)} \left(B\left(\frac{\kappa}{2},\frac{1}{2}\right) - B_{\frac{\kappa}{\kappa+x^2}}\left(\frac{\kappa}{2},\frac{1}{2}\right) \right)$$

$$= 1 - I_{\frac{\kappa}{\kappa+x^2}}\left(\frac{\kappa}{2},\frac{1}{2}\right) = I_{\frac{x^2}{\kappa+x^2}}\left(\frac{1}{2},\frac{\kappa}{2}\right) \tag{5}$$

where; I correspond to incomplete Beta function. Therefore, the search space corresponding to variable $x_{i,j}$ is now divided into $[-1, 0]$ and $[0, 1]$ and instead of applying one cPDF, a pair of cPDFs $P_j(x)$ (6) and $Q_j(x)$ (7) are employed for better sampling based on a parameter ξ that controls the probability of sampling.

$$P_j(x) = \frac{1}{2} - \frac{1}{2} I_{\frac{x_{ij}^2}{\kappa+x_{ij}^2}}\left(\frac{1}{2},\frac{\kappa}{2}\right) \qquad for \quad -1 < x < 0 \tag{6}$$

$$Q_j(x) = \frac{1}{2} + \frac{1}{2} I_{\frac{x_{ij}^2}{\kappa+x_{ij}^2}}\left(\frac{1}{2},\frac{\kappa}{2}\right) \qquad for \quad 0 \le x < 1 \tag{7}$$

These equations are employed to refine the sampling process which ultimately enhance the convergence rate of the proposed metaheuristic globally.

7.4.2 IMPROVED SOLUTION SEARCH EQUATION

Differential evolution (DE) [30] employs the most powerful stochastic real-parameter algorithms to solve multi-modal optimization problems with the optimal combination of population size, and their associated control parameters. In other words, a well-contrive parameter adaptation approach can effectively solve various optimization problems, and convergence rate can improve further if the control parameters are adjusted to appropriate values with improved solution search equations at different evolution stages of a specific problem. There are various DE variants which are different in their mutation strategies, but DE/rand-to-best/1 [31, 33] is one of its kind which explore the *best* solutions to direct the movement of the current population and can effectively maintain population diversity as well.

$$DE/rand-to-best/1: v_t = x_t + SF_1(x_{bes} - x_t) + SF_2(x_r - x_s) \qquad (8)$$

where; SF_1 and SF_2 are scaling factors for neighborhood search. Inspired by this DE variant (8) and inculcating properties of the ABC metaheuristic, we propose a new solution search equations ABC/rand-to-opt/1 as follows:

$$ABC/rand-to-opt/1: v_{ij} = x_{ij} + \varphi_{ij}(x_{opt,j} - x_{ij}) + \psi_{ij}(x_{r1j} - x_{r2j}) \qquad (9)$$

where; r_1, r_2 are random variables from 1, 2, ..., SN, x_{opt} is the optimal individual solution with optimal fitness in the current population with φ_{ij} and ψ_{ij} are scaling factors, respectively.

The proposed solution search equation *ABC/rand-to-opt/1*, which utilizes the information of only optimal solutions in the current population can improve the convergence rate of the proposed metaheuristic.

To increase the multifariousness of the population further, a crossover operation is performed as:

$$u_{ij} = \begin{cases} v_{ij} & \text{if } r[0,1] \le CR, \\ x_{opt,j} & \text{otherwise} \end{cases} \qquad (10)$$

Then a selection operation will be performed as:

$$x_{i,j} = \begin{cases} u_{ij} & \text{if } f(u_{ij}) \leq f(x_{ij}), \\ x_{opt,j} & otherwise \end{cases} \qquad (11)$$

where; $f(x_{ij})$ is the fitness function, if the new solution seems to have high fitness value, then it replaces the corresponding old solution; otherwise, the old solution is retained in the memory. Therefore, with the proposed improved solution search equation, optimal solution is obtained with optimal exploration and exploitation ability thus contribute to a better convergence rate.

7.5   PROPOSED CLUSTERING PROTOCOL-IABC²

Further inheriting the capabilities of the proposed metaheuristic to solve well known multi-modal optimization problem of energy-efficient clustering in UWSNs, an improved Artificial bee colony-based clustering ($iABC^2$) [27, 28] protocol is presented with an optimal CH selection ability.

7.5.1   OPTIMAL CH SELECTION

CH selection is one of the crucial tasks for cluster formation in UWSNs as it affects the overall performance of the network. CH will be responsible for collection of data coming from various SNs and transmission of aggregated data to the BS. Selection of appropriate node as a CH will remain a challenging multi-modal optimization problem. Therefore, we propose an optimal CH selection algorithm based on our proposed *iABC* metaheuristic for an improved energy-efficient clustering protocol. The working of proposed algorithm is as follows.

7.5.1.1   INITIALIZATION PHASE

The population number (PN), corresponding food sources (SN) are initialized along with control parameters Maximum cycle number (MCN), control parameter ξ and Crossover rate (CR).

We employ the proposed improved sampling technique of *iABC* metaheuristic to generate i-th food source $x_{i,j}$, for which we generate $r\varepsilon$ [0, 1] according to uniform distribution and obtain $x_{i,j}$ as:

$$
x_{ij} =, \begin{cases} \dfrac{1}{2} - \dfrac{1}{2} I_{\frac{x_{ij}^2}{\kappa + x_{ij}^2}}\left(\dfrac{1}{2}, \dfrac{\kappa}{2}\right) & \text{if } r \leq \xi, \\[2em] \dfrac{1}{2} + \dfrac{1}{2} I_{\frac{x_{ij}^2}{\kappa + x_{ij}^2}}\left(\dfrac{1}{2}, \dfrac{\kappa}{2}\right) & r \leq \xi, \end{cases} \tag{12}
$$

7.5.1.2 EMPLOYED BEE PHASE

Now each employed bee selects a new solution v_{ij} using proposed improved search Eqn. (10) of proposed *iABC* metaheuristic as:

$$
v_{ij} = x_{ij} + \varphi_{ij}\left(x_{opt,j} - x_{ij}\right) + \psi_{ij}\left(x_{r1j} - x_{r2j}\right) \tag{13}
$$

The obtained value of v_{ij} is compared to x_{ij} and if the fitness of v_{ij} comes out better than x_{ij}, the bee will forget the previous old solution and retain the new optimal solution $x_{opt,j}$ found so far, otherwise, it will keep working on x_{ij}.

7.5.1.3 ONLOOKER BEE PHASE

Now, employee bee will share the information of their food source with the onlooker bee, through a *waggle* dance performed at their hive, each of whom will then generate a food source u_{ij} according to distribution as:

$$
u_{ij} = \begin{cases} v_{ij} & \text{if } r[0,1] \leq CR, \\ x_{opt,j} & otherwise \end{cases} \tag{14}
$$

where; CR is crossover rate, further fitness of generated food source $f(u_{ij})$ is calculated and compared with previous food source as:

$$
x_{i,j} = \begin{cases} u_{ij} & \text{if } f(u_{ij}) \leq f(x_{ij}), \\ x_{opt,j} & otherwise \end{cases} \tag{15}
$$

where; $f(x_{ij})$ is the fitness value of x_{ij}. Onlooker bee will then choose a food source with higher fitness and conduct a local search on x_{ij}, if the new solution has a better fitness, then it will replace x_{ij} with optimal solution $x_{opt,j}$ and assigned as a CH, otherwise the old solution will be retained.

7.5.1.4   SCOUT BEE PHASE

Now, if the fitness cannot improve further, after a number of trials, then the corresponding employed bee becomes a scout to produce a new food source randomly by using Eqn. (12) again.

The detail Cluster Head (CH) Selection Algorithm is discussed as below:

```
Optimal Cluster Head(CH) Selection Algorithm
Input:
PN ← Population number.
MCN ← Maximum cycle number,
D ← Dimension of vector to be optimized.
SN ← Food sources.
x_min ← Lower bound of each element.
x_max ← Upper bound of each element.
ξ ← Control parameter.
CR ← Crossover rate.
Output:
Ch_j ← x_opt,j

begin
round ← 0
for i = 1 → SN do
    Generate rε [0,1] according to uniform distribution.
    if r ≤ ξ then
        Generate x_ij ε [−1,0] according to PDF P_j(x).
    else
        Generate x_ij ε [0,1] according to PDF Q_j(x).
        Evaluate fitness f_i(x_ij)
        trial(s) ← 0
        round ++
    end if
end for
repeat
until
for i = 1 → SN do
    Generate v_ij according to Eq.
    Evaluate fitness f_i(v_ij)
    round ++
    if f_i(x_ij) < f_i(v_ij) then
        x_ij ← v_ij
        f_i(x_ij) ← f_i(v_ij)
        trial(s) ← 0
    else
        trial(s) ← trial(s) + 1
    end if
end for
if round == MCN then
    Memorize the optimal soloution, x_opt,j achieved so far and exit repeat.
    Ch_j ← x_opt,j
end if
repeat
until
for i = 1 → SN do
    r ← rand[0,1]
    if r ≤ CR then
        u_ij ← v_ij
    else
        u_ij ← x_opt,j
    end if
    Evaluate fitness f_i(u_ij) and f_i(x_opt,j)
    if f_i(u_ij) ≤ f_i(x_opt,j) then
        x_i,j ← u_ij
        if f_i(u_ij) > f_i(x_opt,j) then
            x_opt,j ← u_ij
            f_i(x_opt,j) ← f_i(u_ij)
            trial(s) ← trial(s) + 1
        end if
    end if
    if solution need to be abandoned
    replace with a new solution, produced using Eq.(12)
    round ++
end for
if round == MCN then
    Memorize the optimal soloution, x_opt,j achieved.
    Ch_j ← x_opt,j
end if
```

7.6   RESULTS AND DISCUSSIONS

Now we evaluate the performance of the proposed protocol with existing PSO [15], MRP [35], and LEACH [10] protocols using ns-2 simulator. The protocols are simulated over two different BS position scenarios to assess their behavior towards energy consumption, network lifetime and end-to-end delay. The simulation will be performed over standard MAC protocol with Free Space radio propagation and CBR traffic type, considering other parameters as shown in Table 7.1.

TABLE 7.1 Simulation Parameters

Parameter	Value
Terrain size	150×150 m^2
MAC protocol	802.11
Radio propagation	Free space
Traffic type	CBR
ε_{fs}	6 pJ/bit/m
ε_{mp}	0.0011 pJ/bit/m^4
Propagation limit	–111 dBm
Receiver sensitivity	–89
Data rate	2 Mbps
Packet size	3,000 bits
Message size	400 bits

In the first scenario UWSN #1, a network of sensor nodes ranging from 75 to 450 are deployed randomly over an area of size 150×150 m^2 with a BS, located at position (50 m, 100 m) within the network field whereas in second scenario UWSN #2, a BS will be placed at position (100 m, 200 m) outside the network field. First, we execute the protocols to compare energy consumption in the network for both the scenarios.

Figure 7.1 shows that in scenario UWSN #1, energy consumption of the proposed protocol *iABC*2 is approximately 26%, 61%, 123% less then PSO, MRP, and LEACH protocols, respectively, which is attributed to the use of compact student's-t distribution and improved solution search equation to selects optimal CHs, thus minimize energy consumption in the network. Even in scenario UWSN #2 (Figure 7.2), the proposed protocol consumes 33% less energy as compared to its contender PSO, which clearly shows the effectiveness of the proposed metaheuristic *iABC*. In *iABC*2, Optimal CHs are selected not only based on their proximity to BS but with the condition of minimum power consumption in data transmission, moreover the SNs are assigned to their nearest CH, thus consume less energy and as a result the overall energy consumption of the network becomes lesser than other protocols. In PSO, all CHs are inevitably used as a relay node to forward the data packets to the BS, therefore consume more energy, whereas MRP and LEACH selects CHs based on a pre-defined selective probability of sensor nodes which will not optimize the energy usage in complex UWSNs.

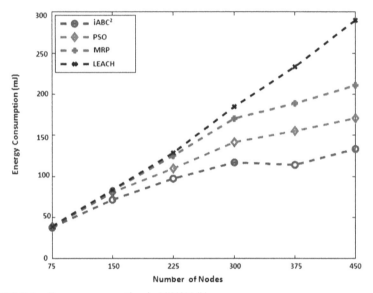

FIGURE 7.1   Energy consumption in UWSN #1.

Further, Figures 7.3 and 7.4 show that *iABC*2 extend the average network lifetime by approximately 18% and 12% compared to MRP and PSO in UWSN #1 and UWSN #2 respectively, which is the outcome of nodes surplus energy availability due to less computation coupled with better convergence and an optimal selection of CHs with proposed metaheuristic. The energy thus saved in *iABC*2 will prolong the network lifetime and the nodes will be able to transmit data for a longer duration. In PSO, due to un-symmetric data forwarding effects on the CHs, those near to the BS will die quickly thus reduces network lifetime. LEACH is having the least network lifetime among all its peers, due to absence of a clear data aggregation and communication framework, especially for UWSN # 2 like scenarios. It is further analyzed that every 1% increase in network lifetime of the proposed protocol, will increase the data delivery by 3.2%, thus increasing networks robustness.

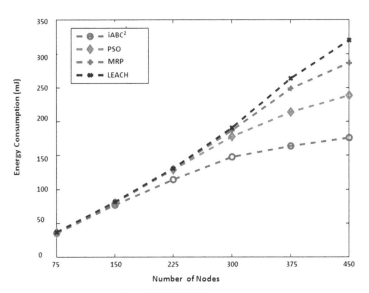

FIGURE 7.2   Energy consumption in UWSN #2.

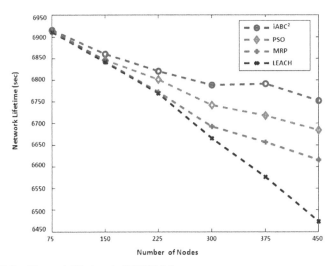

FIGURE 7.3   Network lifetime in UWSN # 1.

Figures 7.5 and 7.6 compares the end-to-end delay in both scenarios after number of rounds ranging from 100 to 600. It is clearly visible that *iABC*2 deliver data packets with minimum end-to-end delay in both the scenarios among other protocols, which ultimately increase the reliability of the network. In UWSN #1, end-to-end delay decreases sharply with increase in number of rounds in *iABC2*, which is due to the fact that the proposed protocol delivers data packets to the BS with minimum relay after calculating the optimal possible distance for the next hop, moreover the CHs are placed at optimal distance to BS thus maintains a trade-off between transmission distance and hop-count. Also, in UWSN #2, when the BS is located at a far distance from sensor nodes, the proposed protocol will be able to deliver the data packets with minimum delay successfully. In PSO and MRP, data will be transmitted to BS using maximum number of hop-count ultimately exhaust the network with unnecessary end-to-end delay.

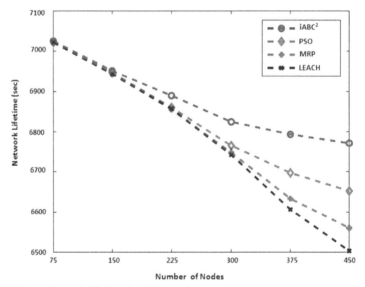

FIGURE 7.4   Network lifetime in UWSN # 2.

7.7   CONCLUSION

This chapter presents an *iABC* metaheuristic, which is based on first of its kind student's-t cPDF and DE inspired improved solution search equation to improve exploitation capabilities as well as convergence rate of existing ABC metaheuristic. Further, we presented *iABC2*, a clustering protocol based on the proposed metaheuristic for UWSNs, which selects optimal

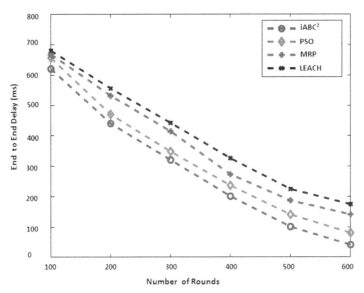

FIGURE 7.5   End-to-end delay in UWSN # 1.

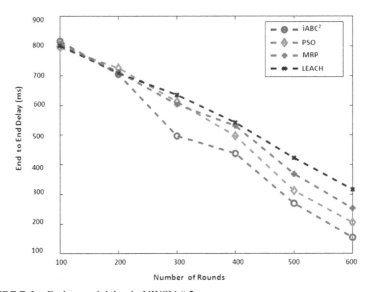

FIGURE 7.6   End-to-end delay in UWSN # 2.

CHs based on an improved search equation and an efficient fitness function. At last, we compare the performance of the proposed protocol with other protocols to prove its validness over various performance metrics.

Simulation results show that *iABC*2 consumes approximately 26% to 123% less energy than PSO, MRP, and LEACH protocols and prolong network life approximately by 12% to 18% with minimum end-to-end delay in diverse UWSNs scenarios.

KEYWORDS

- **artificial bee colony**
- **clustering protocol**
- **computational intelligence**
- **internet of underground things**
- **underground wireless sensor networks**

REFERENCES

1. Akkaya, K., & Younis, M., (2005). A survey on routing protocols for wireless sensor networks. *Ad Hoc Networks, 3*(3), 325–349.
2. Al-Karaki, J. N., & Kamal, A. E., (2004). Routing techniques in wireless sensor networks: A survey. *Wireless Communications, 11*(6), 6–28. IEEE.
3. Attea, B. A., & Khalil, E. A., (2012). A new evolutionary based routing protocol for clustered heterogeneous wireless sensor networks. *Applied Soft Computing.*
4. Bayrakdar, M. E., (2020). Cooperative communication-based access technique for sensor networks. *International Journal of Electronics, 107.*
5. Chamam, A., & Pierre, S., (2010). A distributed energy-efficient clustering protocol for wireless sensor networks. *Computers & Electrical Engineering, 36*(2), 303–312.
6. Kandris, D., Nakas, C., & Vomvas, D., (2020). Applications of wireless sensor networks: An up-to-date survey. *Applied System Innovation, 3.*
7. Deng, S., Li, J., & Shen, L., (2011). Mobility-based clustering protocol for wireless sensor networks with mobile nodes. *Wireless Sensor Systems, IET, 1*(1), 39–47.
8. Mininno, E., Cupertino, F., & Naso, D., (2008). Real-valued compact genetic algorithms for embedded microcontroller optimization. *IEEE Trans. Evol. Computer, 12*(2), 203–219.
9. Gaura, E., (2010). *Wireless Sensor Networks: Deployments and Design Frameworks.* Springer.
10. Heinzelman, W. B., Chandrakasan, A. P., Balakrishnan, H., et al., (2002). An application-specific protocol architecture for wireless microsensor networks. *IEEE Transactions on Wireless Communications, 1*(4), 660–670.

11. Hoang, D., Yadav, P., Kumar, R., & Panda, S., (2014). Real-time implementation of a harmony search algorithm-based clustering protocol for energy efficient wireless sensor networks. *IEEE Transactions on Industrial Informatics*.

12. Jin, Y., Wang, L., Kim, Y., & Yang, X., (2008). Eemc: An energy-efficient multilevel clustering algorithm for large-scale wireless sensor networks. *Computer Networks, 52*(3), 542–562.

13. Karaboga, D., & Akay, B., (2009). A comparative study of artificial bee colony algorithm. *Applied Mathematics and Computation, 214*(1), 108–132.

14. Karaboga, D., & Basturk, B., (2008). On the performance of artificial bee colony (ABC) algorithm. *Applied Soft Computing, 8*(1), 687–697.

15. Kuila, P., & Jana, P. K., (2014). Energy-efficient clustering and routing algorithms for wireless sensor networks: Particle swarm optimization approach. *Engineering Applications of Artificial Intelligence, 33*, 127–140.

16. Kulkarni, R. V., Forster, A., & Venayagamoorthy, G. K., (2011). Computational intelligence in wireless sensor networks: A survey. *Communications Surveys & Tutorials, IEEE, 13*(1), 68–96.

17. Kumar, D., Aseri, T. C., & Patel, R., (2009). EEHC: Energy efficient heterogeneous clustered scheme for wireless sensor networks. *Computer Communications, 32*(4), 662–667.

18. Samrat, L., Udgata, S., & A. A., (2010). Artificial bee colony algorithm for small signal model parameter extraction of MESFET. *Eng. Appl. Artif. Intelligence, 11*, 1573–1592.

19. Liu, Z., Zheng, Q., Xue, L., & Guan, X., (2012). A distributed energy-efficient clustering algorithm with improved coverage in wireless sensor networks. *Future Generation Computer Systems, 28*(5), 780–790.

20. Mao, S., Zhao, W., & Cheng-Lin, (2011). Unequal clustering algorithm for WSN based on fuzzy logic and improved ACO. *The Journal of China Universities of Posts and Telecommunications, 18*(6), 89–97.

21. Larranaga, P., & Lozano, J., (2001). *Estimation of Distribution Algorithms: A New Tool for Evolutionary Computation*. Kluwer.

22. Palvinder, S. M., & Singh, S., (2017a). Artificial bee colony metaheuristic for energy-efficient clustering and routing in wireless sensor networks. *Soft Computing, 21*, 6699–6712. Springer.

23. Palvinder, S. M., & Singh, S., (2017b). Energy-efficient clustering protocol based on improved metaheuristic in wireless sensor networks. *Journal of Network and Computer Applications, 83*, 40–52. Elsevier.

24. Palvinder, S. M., & Singh, S., (2017c). Energy-efficient clustering protocol based on improved metaheuristic in wireless sensor networks. *Journal of Network and Computer Applications, 83*, 40–52. Elsevier.

25. Palvinder, S. M., & Singh, S., (2017d). Improved metaheuristic-based energy-efficient clustering protocol for wireless sensor networks. *Engineering Applications of Artificial Intelligence, 57*, 142–152. Elsevier.

26. Palvinder, S. M., & Singh, S., (2019a). Improved artificial bee colony metaheuristic for energy-efficient clustering in wireless sensor networks. *Artificial Intelligence Review, 51*, 329–354. Springer.

27. Palvinder, S. M., & Singh, S., (2019b). Improved metaheuristic-based energy-efficient clustering protocol with optimal base station location in wireless sensor networks. *Soft Computing, 23*, 1021–1037. Springer.

28. Das, S., Abraham, A., & Konar, A., (2009). Metaheuristic clustering. *Studies in Computational Intelligence*, 178.

29. Selvakennedy, S., Sinnappan, S., & Shang, Y., (2007). A biologically-inspired clustering protocol for wireless sensor networks. *Computer Communications, 30*(14), 2786–2801.

30. Storn, R. P. K., (2010). Differential evolution-a simple and efficient heuristic for global optimization over continuous spaces. *Journal of Global Optimization, 23*, 689–694.

31. Swagatam, D., & Suganthan, P. N., (2011). Differential evolution: A survey of the state-of-the-art. *IEEE Transactions on Evolutionary Computation*, 15.

32. Tyagi, S., & Kumar, N., (2012). A systematic review on clustering and routing techniques based upon leach protocol for wireless sensor networks. *Journal of Network and Computer Applications, 12*, 92–110.

33. Gonuguntla, V., Mallipeddi, R., & Veluvolu, K. C., (2015). Differential evolution with population and strategy parameter adaptation. *Mathematical Problems in Engineering, 78*, 145–160.

34. Walck, C., (2007). *Statistical Distributions for experimentalists*. Particle Physics Group.

35. Yang, J., Xu, M., Zhao, W., & Xu, B., (2009). A multipath routing protocol based on clustering and ant colony optimization for wireless sensor networks. *Sensors, 10*(5), 4521–4540.

36. Yi, S., Heo, J., Cho, Y., & Hong, J., (2007). Peach: Power-efficient and adaptive clustering hierarchy protocol for wireless sensor networks. *Computer Communications, 30*(14), 2842–2852.

37. Yick, J., Mukherjee, B., & Ghosal, D., (2008). Wireless sensor network survey. *Computer Networks, 52*(12), 2292–2330.

38. Younis, O., & Fahmy, S., (2004). Heed: A hybrid, energy-efficient, distributed clustering approach for ad hoc sensor networks. *IEEE Transactions on Mobile Computing, 3*(4), 366–379.

39. Zhang, R., & Wu, C., (2011). An artificial bee colony algorithm for the job shop scheduling problem with random processing times. *Entropy, 13*(9), 1708–1729.

CHAPTER 8

Smart Industry Pollution Monitoring and Control Using the Internet of Things

KUNJABIHARI SWAIN,[1] AMIYA RANJAN SENAPATI,[2]
SANTAMANYU GUJARI,[3] and MURTHY CHERUKURI[1]

[1]*Department of Electrical and Electronics Engineering, National Institute of Science and Technology, Berhampur, Odisha–761008, India, E-mail: kunja.swain@gmail.com (K. Swain)*

[2]*Project Engineer, Wipro Technologies, Karnataka–560035, India*

[3]*Sensor Specialist, KONE Elevator India Pvt. Ltd., Maharashtra–411001, India*

8.1 INTRODUCTION

IoT is commonly referred to as the Internet of Things, is a nexus of practical devices and objects affiliated by the modern technology of connectivity through various networks that facilitate them to collect and share the data over the internet. With the industrial revolution, the expansion of industries in all the sector around the world have cumulatively resulted in industrial pollution inducing climatic and biological changes spawning annihilation of the ecosystem. In order to reduce the wide range of industrial pollution effect on the biodiversity, several practices are being followed. Treatment of the waste products produced by the industries are crucial to control and minimize the effect that is caused by the industrial effluents. Physical, chemical, biological treatment are the way which are implemented to overcome and minimize this crisis. With the advent of new technologies

Harnessing the Internet of Things (IoT) for Hyper-Connected Smart World.
Indu Bala and Kiran Ahuja (Eds.)

and advancement in the networks, human being has always come up with a new strategy to mitigate various hazards. The motivation of this chapter is to provide industrial pollution control practices using IoT-based approach. This chapter proposes a smart waste management system to determine the maximum pollution load (via monitoring section) and achieve an average permissible point which could be used by the plant operators to vary and control (via controlling section) the quality standards of the industrial waste prior discharging to the environment.

Currently there are numerous IoT-based models which were used for various applications such as in creating smart home wherein the microcontrollers are used in combination with various sensors to control various range of electrical equipment such as television, refrigerator, air conditioners at home. In addition to this controlling, there are smart home solutions where the temperature, humidity, and motions are also captured for surveillance [1, 2]. IoT-based models are also used in tracking energy consumptions in any environmental setups in school premises, laboratories, etc. The power consumptions of the devices are scrutinized and monitored and based on the consumption, curative actions are taken to reduce energy consumption [3]. In water quality monitoring system, IoT-based models are used to measure various physicochemical parameters of water to detect the water contamination level and help in providing clean water in real-time [4]. The ambient temperature and humidity within a storage industry play a very vital role for sustaining the products quality, IoT models are used in such setup to have an acute watch on these parameters, specially in food storage industries [5]. In Ref. [6], the author demonstrated the establishment of a basic IoT system inside an industry to enhance the safety level of the worker using LabVIEW and Arduino microcontroller. There are several application protocols and technology proposed for IoT solution [7–10]. Reference [11] presents different architecture and key technology of IoT. In Ref. [12], IoT-based hierarchical architecture has been proposed for the smart factory. In Ref. [13], an IoT-based dynamics production logistics synchronization of manufacturer presented. Different security issues of IoT are presented in Ref. [14]. Ref. [15] presents a comparison of different WLAN technologies which can be used for IoT. In Refs. [16–18], the authors discussed industrial IoT challenges, opportunities, design patterns and the potential research directions in the industries. A smart factory framework is presented in Ref. [19] to provide cloud service to the industrial network. In Refs. [20, 21], the authors presented a smart city and application based on IoT.

This chapter presents an IoT-based system to monitor the permissible pollution load that is safe enough for a plant prior to disposing to the environment and have a keen check on the operational parameters of the effluent standards by varying the control actuators. This will ensure in keeping the environment safe and free from pollution which in turn will reduce the various biohazards affecting billions of life in this ecosystem. It also assures the quality of the input ingredient such as water or any other variable needed in the plant for daily usage.

The remainder of the chapter is categorized as follows. Section 8.2 explains the methodology adopted for the proposed work. Section 8.3 describes the working and the case study. Section 8.4 presents the results. Finally, the concluding remarks are provided in the last section.

8.2 METHODOLOGY

The system proposed in this chapter emphases on the monitoring the different industrial parameters like the temperature, humidity, fire detection, pH, content of carbon dioxide, detection of smoke, detection of fire, etc. The system would also be capable of controlling suitable protection elements to reduce the hazard and activate the alarming system through an android application, to provide IoT-based interactive monitoring and efficient control. To realize the proposed idea, different sensors are installed at suitable locations of the industries, based on the Wi-Fi ranges of operation of the sensors. The sensor data are collected by the esp8266 Wi-Fi modules and then communicated to the control room. Additionally, a IoT-based interactive monitoring and control feature was implemented to enhance the functionality and performance using android app. Figure 8.1 shows the proposed methodology.

A database is setup in the local machine which is configured to store the sensor data generated during the operation, by using the SQL queries. In this present implementation, as the client and server remain in the same local machine, it is regarded as the local host, i.e., the parent local machine is treated as the webserver. For the overall operation, a database named "IoT Database" is created with several tables such as "temperature sensor," "pH sensor," and "humidity sensor", which stores the data generated from the individual sensors. The tables in the database enables to study and manage the individual data and analyzing them.

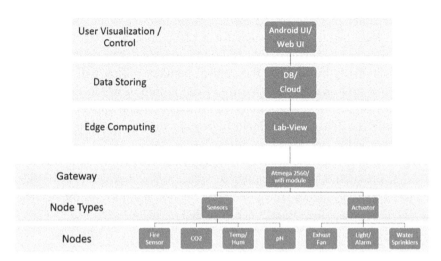

FIGURE 8.1   Block diagram of the proposed technique.

8.3   WORKING AND CASE STUDY

8.3.1   SOFTWARE

The software architecture of the overall proposed system is implemented by using a web server solution stack called XAMPP.

XAMPP is an acronym for XAMPP Apache, Maria DB, PHP, and PERL. The following sections briefly describe the main components of XAMPP.

8.3.1.1   APACHE HTTP SERVER

Apache HTTP server is commonly available software, which acts as a middleware in between the client, server web architecture. It acts as a middleman between client and server machine, by helping in exchanging the information back and forth between both the systems.

The Apache is software that runs in the HTTP server which is used to establish the connection between the server and the browsers (such as Chrome, Safari, Mozilla, and Internet Explorer, etc.). The HTTP server hosts the website in the local machine. The website is a user interface used for monitoring and controlling and described in brief in the website section.

8.3.1.2 DATABASE

The key component of any IoT-based approach is uniquely its data which is generated from the various hardware components involved in the overall operation. The efficient management of this unique data in an IoT-based model indeed depends on the collective approach which is stored in a database. The profitable approach is to use a relational database, and it has been very much effective in managing many IoT-based models in the market. Some of the factors that influence in choosing the database are such as size, scale, amount of data, query language, etc. The XAMPP solution stack provides a database called MARIADB. It is a freely available RDBMS (Relational Database Management System), which resembles to MYSQL in terms of operation and compatibility. It uses Structured Query Language (SQL) to create control and modify the data stored in the database.

8.3.1.3 PHP SCRIPTING

PHP is a scripting language commonly used in web server architecture, comes handy with an PHP interpreter which serves as the core of the PHP language. Usually, the interpreter processes the code and implement the necessary actions and gather the response to/from the webserver. In this model, the PHP code is implemented in such a way that it connects to the "IoT Database" and pushes different sensor values into different tables.

Figure 8.2 shows a sample PHP code which is used to connect to the database.

```php
<?php
$conn = new mysqli('localhost', 'root', '', 'mydb');
if ($conn->connect_error) {
    die("Connection error: " . $conn->connect_error);
} ?>
```

FIGURE 8.2   A sample PHP code.

Once the connection is established with the database (IoT Database), the above code as shown in Figure 8.2, returns a connection variable "$conn." Based on this connection variable, various operations are performed on the tables present in the IoT Database. The sensor data is stored in tables continuously. The PHP scripts implemented with the HTML code, which in turn accumulate the equivalent data and shows in the MONITORING section of the webpage. The LabVIEW software is used to perform the data inserting and retrieving to/from the database.

8.3.1.4   WEB APPLICATION

A web-based application is built using HTML5 and PHP script. This application takes data from the 'IoT Database' tables and displays it on the webpage. The web application has two tabs under it as follows:

1. **Monitoring:** The monitoring section of the web application is designed to preview the respective data from the pH, humidity, and temperature sensors which are pushed into the corresponding tables of the database. For surveillance, this section of the website provides a hyperlink which will redirect the operator to a page which would show the live CCTV footage of the plant.

2. **Controlling:** Based on the continuous live feed shown in the monitoring section, the operator can control various control parameters. Button switches present in the controlling section enables the user to control different parameters. For demonstrating temperature and humidity control, a fan and a heater are connected with the button switches. By turning on and turning off these switch buttons for the specified time, the temperature and humidity can be brought to normal range. This is essential to have a keen watch over the pollution parameters generated from an industry during its operation. For 360° surveillance, the rotation of CCTV camera is controlled with the buttons present in this section. The controlling and monitoring section of the web application is built using PHP scripts. Under the controlling tab, upon operators input, predefined Boolean values are pushed into the "Control" table of the database, and after every refresh of the feed appropriate action are triggered by the Arduino board.

8.3.1.5   ANDROID APPLICATION

The Android application is built in the Android studio to improve performance and functionality. The Android application consists of two major parts, i.e., monitoring and control. The monitoring section displays the humidity, temperature, and pH values taken from the respective tables of the "IoT Database." The control section comprises of switch buttons to operate different actuator like the fan, the heater, and the chemical valves to bring the parameters within the safe range. PHP scripts are run by the android application to retrieve and insert data from and to the tables.

8.3.1.6   LabVIEW

LabVIEW is a graphical programming language with a graphical interface for easy and fast coding. National Instruments allows a number of third-party devices to connect with LabVIEW program, which includes the commonly used Arduino microcontroller. The VISA (Virtual Instrument Software Architecture) Sub VI plays a vital role in LabVIEW, which allows us to connect to the Arduino board over serial/COM port of the personal computer. LabVIEW is used to obtain pH, humidity, temperature sensor data via Arduino board. Upon gathering all the sensor data, they are stored in an appropriate table of the database.

The procedure starts with configuring the Arduino microcontroller. The microcontroller is coded in such a way that it sends the sensor data when it receives a serial request command. The serial input is given to the microcontroller from LabVIEW through the serial/COM port. With reference to the requested command, the Arduino board responds back the sensor data, the respective data are stored in different database tables using MySQL Sub VI's.

8.3.2   HARDWARE

The hardware sections consist of various sensors to sense various parameters within an industry and send to the gateway. The actuators receive the command from the gateway and operate the load. The various sensors used are described as follows.

8.3.2.1 pH SENSOR

To measure the acidity of disposed waste liquid coming out from the plant, we have a pH sensor. The pH electrode has an output range of 400 mV to

−400 mV for the pH value of 1 to 14. To fetch the pH sensor signal, we need to use a signal conditioning circuit. This will add an offset value in the raw sensor data. Hence the sensor ranges of −400 mV to 400 mV can be mapped to a range of 500 mV to 1,300 mV. This output range can directly feed into any controller. In this case, the Wi-Fi module, i.e., esp8266 is used as a controller to acquire the pH data and forward it to the gateway.

The pH value is dependent on the temperature factor, so we can use the below equation to calculate the compensated pH value.

$$pHC = pH - ((T - T0) \times (pH0 - pH) \times 0.003)$$

where; pHC is the compensated pH value; pH is the measured pH value; pH0 is the center pH value of 7; T is the temperature in degree Celsius; T0 is the center temperature value of 25°C; 0.03 is the correction factor in pH/°C/Ph.

8.3.2.2   DHT11

It is essential to monitor the temperature and humidity level for the safe-guard of daily worker and produced goods. DHT 11 sensor is employed to monitor both temperature and humidity. This does not require additional signal condition circuit or calibration.

DHT 11 sensor module has a humidity sensing element which is of capacitive type and a thermistor to sense the temperature. The humidity sensor works as a capacitor. It has a moisture holding substrate, which acts as a dielectric material of a capacitor. With the variation in humidity level, the capacitance is also modified. It is further connected to an IC to measure and process the change in capacitance value into digital form. It uses a thermistor of negative temperature coefficient type to measure tempera-ture, where resistance variation is inversely proportional to temperature. The module has four terminal-VCC, Data, GND, and a no use pin. The sensor communicates with the micro-controller through digital signal, which prevent unnecessary fluctuations.

8.3.2.3   MG-811

This sensor working is based on the principle of a solid electrolyte cell. Whenever the sensor is exposed to CO_2, different electrode reactions occur and the EMF generated is based on the Nernst equation. When the maximum surface temperature level is reached, the sensor acts as a single cell generating output voltage signal from its two terminals. The change in EMF output signal of the module is 25 mV/1,000 ppm of CO_2. This signal can be directly feed to the analog pin of a controller. In this, we will feed to the esp8266 Wi-Fi module, which acts as a node [22]. This will send the sensor data to the gateway. Figure 8.3 shows the CO_2 sensor module.

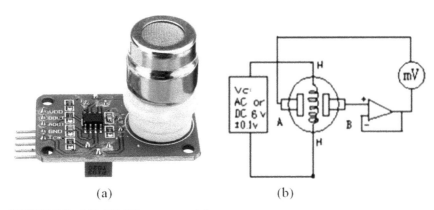

(a)   (b)

FIGURE 8.3 MG-811 CO_2 sensor module.

When the sensor meets the CO_2 particles, certain electrode reaction takes place within the sensor chamber. The reaction can be expressed as:

- **Reaction at Cathode Terminal:** $2Li+ + (0.5)O_2 + CO_2 + 2e^- = Li_2CO_3$
- **Reaction at Anode Terminal:** $(0.5)O_2 + 2Na+ + 2e^- = Na_2O$
- **Overall Reaction:** $2Na++Li_2CO_3 = 2Li++CO_2+Na_2O$

According to Nernst's equation, the electromotive force (EMF) resulted from the above electrode reaction is:

$$EMF = Econ - ((R \times Ta)/(2F)) \times \ln(P(CO_2))$$

where; Econ is the constant volume; R is the gas constant volume; Ta is the absolute temperature (Kelvin); F is the Faraday constant; $P(CO_2)$ is the partial pressure of CO_2.

The heating voltage of the sensor provided from the circuits. when the surface temperature is reasonably high, is the same as that of the battery, the two sides of the sensor provide the output voltage signal and the affect the Nernst equation. The amplifier impedance has to be within 100–1,000 GΩ, which can amplify the signal with current less than 1 pA.

8.3.2.4   SMOKE SENSOR

In case of fire outbreak, it can detect early and save the life and property. The sensor is based on photoelectric, or optical smoke detector technology, as shown in Figure 8.4. This is cost-effective and easily installed because of wireless technology. The node esp8266 will send the data to the gateway esp8266 module.

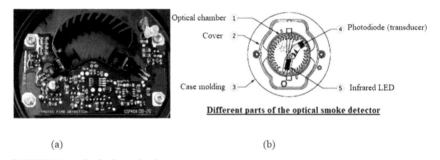

Different parts of the optical smoke detector

(a)   (b)

FIGURE 8.4   Optical smoke detector.

It uses photoelectric technology, where an IR transmitter and a receiver are placed within the chamber. It is placed in such a way that the detector is not perpendicular to the IR transmitter. During the fire outbreak, the smoke particles enter the optical chamber through a small windows of the chamber. Then the IR lights gets diffracted by the smoke particles towards the detector. These transmitter and receiver are connected to an IC. This IC manages the power consumption for maximum battery life, signal amplification, setting threshold to trigger alarm and many more. This alarm digital signal can be fed to a controller to process further [23].

8.3.2.5   IR FLAME SENSOR MODULE DETECTOR

To detect the fire in the plant, a flame sensor is used. The angle of detection is more than 120°, being sensitive to the spectrum of flames. It detects the presence of fire or flame within a premises. Figure 8.5 shows a flame sensor employed in this work [24].

FIGURE 8.5 Flame detector.

8.3.2.6 ESP8266 WI-FI MODULE

ESP8266 is a low powered Wi-Fi chip used for acquiring all the nodes data and forwards all the collected data to the main controller. In this work, ESP8266 module works in hybrid mode. It acts as an access point as well as station, simultaneously. It uses the mesh network as shown in Figure 8.6. The mesh network was built with its own Wi-Fi. Therefore, it does not depend on the factory's existing networking structure. Hence the traffic remains the same. It does not use on TCP/IP protocol. Therefore, it cannot be accessed by hackers remotely and it is easy to implement. Each node is given a unique identification number of 32-bit length, which can be considered as chip ID. Messages from node can either be sent to the gateway through a chain of nodes or broadcasted to all the nodes. Sender can specify the destination with its 'node ID.'

In this implementation, the data communication is done over a long-range. In case of failure of any node in a chain, other nodes act as backup to complete the communication [25].

8.3.2.7 ARDUINO MICROCONTROLLER

Arduino AT mega microcontroller is used for connecting the wired sensors and the ESP8266 wireless module, which collect the sensor data from different sensors. It works as a central unit and communicates to the monitoring/controlling sections.

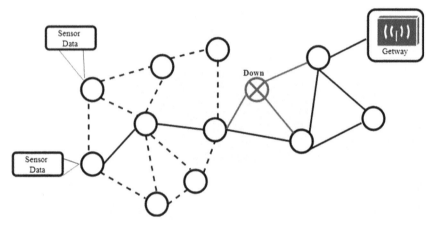

FIGURE 8.6   Mesh network.

8.3.3   USER VISUALIZATION/CONTROL

The operator can track the various parameters of the industry through a web browser or smartphone and take the required measures in the same network, if required in an emergency situation.

8.3.4   DATA STORING

All sensor data is stored on a local database, which can be accessed through a web application or android application connected on the same network. And all emergency alerts are sent to cloud over MQTT protocol which can be accessed globally over a wide area network.

8.3.5   EDGE COMPUTING

This computing method optimizes hardware, IoT devices and applications by getting computation closer to the edge of those data-producing objects in the network. The edge of the network may apply to the area where the device communicates with the internet, depending on the system or equipment being considered. In this case, we have used a PC as an edge computing device, on which LabVIEW is running.

8.3.5.1 GATEWAY

This acts as a brain of the system, which can take immediate action for any failure and retransmit the data to LabVIEW. It can collect the sensor node data through mesh network topology. It is implemented with the help of an open-source library "*painless Mesh.*" It is an ad-hoc network.

The operation of the proposed working model consists of the two main sections:

1. **Monitoring Section:** DHT-11 humidity and temperature sensor, pH sensor, CO_2 sensor module acts as a node in this network. They forward the data to the main ESP8266 module which is connected to the main Arduino board (Atmega2560), which we have considered as a Gateway in this network. The Arduino board begins sending the data to the PC through a COM/serial port, where LabView software is running. Upon receiving the data, the Arduino serial monitor displays the data. The serial data is feed to the LabVIEW software for interacting with the Gateway. The commands such as "H" for humidity, "T" for temperature and "P" for pH value is sent periodically to the Arduino board to fetch the respective sensor data from the Gateway into the LabVIEW software through virtual instrument software architecture (VISA).

   The data collected in the LabVIEW software is further directed to the corresponding tables such as "Temperature," "Humidity," "pH" tables in the "IoT Database" using LabVIEW code.

   For the web application to be operational, the XAMPP server is started, and upon hitting the localhost URL, the web application is accessible. The objective of the web application is just to fetch its live information feed from the different tables of "IoT Database" and display it in the predefined columns of the monitoring section. The android application is interfaced to the same localhost URL by pairing the mobile device through wi-fi module. The mobile application then registers with the system and get access to the "IoT Database" and fetches the data from the respective tables and displays it in the mobile application.

2. **Controlling Section:** The controlling section of the web application provides a rich set of buttons, generally acting as actuators, performs the predefined operations upon receiving the user inputs. An operator can start the fan, light, chiller, acid valve, base valve by clicking the individual buttons. When we toggle a button, the PHP script feeds the "Control" table with the predefined values. The table is further fed into to Arduino board by LabVIEW. Arduino board upon receiving the serial data from LabVIEW operates the devices connected to it.

The overall working can be best understood from Figure 8.7.

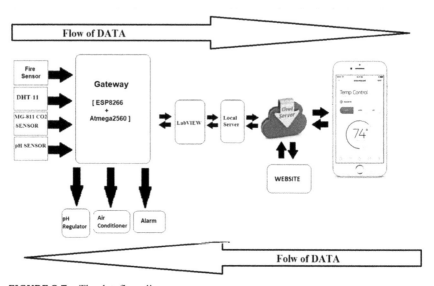

FIGURE 8.7   The dataflow diagram.

There are bi-directional flows of data in the proposed system. For the monitoring section, the data flows from left to right and for controlling section, the flow is from right to left, from the sensors to the monitoring section. After successful integration of various hardware modules with the Arduino board and ESP8266, the system begins with the Arduino board fetching the serial data from individual sensors and further the data is fed into the LabVIEW by using VISA tool. The LabVIEW analyzes the data and after post manipulation, it feeds various tables in the "IoT Database." The local server is hosted on a cloud web hosting. The value present in the cloud server is apparently displayed in the monitoring section of the web and android application.

For controlling of the various actuators, the flow of commands is from right side to the left side. The control inputs are given in the form of clicks to the various buttons present in the controlling panel of the web and android application. As mentioned earlier upon user inputs, the respective PHP scripts are triggered, and predefined values are fed into the table whose name is "Control." LabVIEW forward them to the Serial/COM port, and they are decoded sequentially by the Arduino board which actuates the chillers, fans, control valves and other actuators.

8.4   RESULTS AND DISCUSSION

The data collected from the various sensors are displayed in the monitoring section. An android app was developed to present this information in an android mobile or in a tablet, as shown in Figure 8.8. The same parameters are also presented in the web browser to make the monitoring and control flexible in any platform. The web browser is shown in Figures 8.9 and 8.10. Figure 8.11 shows the LabVIEW front panel showing the monitoring parameters.

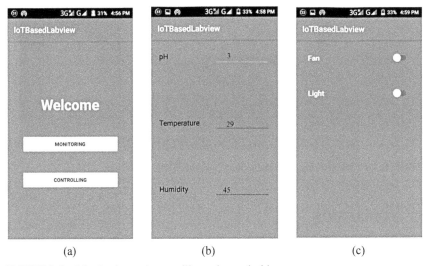

(a)   (b)   (c)

FIGURE 8.8   Monitoring and controlling using android app.

FIGURE 8.9 An overview of the web browser.

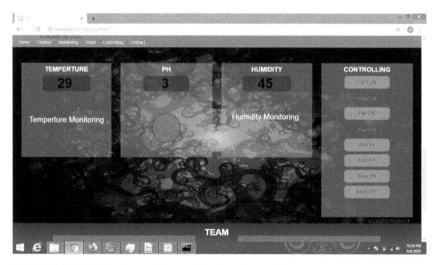

FIGURE 8.10 Monitoring and controlling from the web browser.

8.5 CONCLUSION

The IoT approach may be applied in a variety of applications. The implementation of IoT in controlling and monitoring of industrial pollution has been presented in this chapter. With this implementation, it can provide the safety of the plant workers and the protection of the industrial environment. It also prevents any devastating hazardous waste coming out from the industry as well as provides a certain degree of control over environmental pollution, which affects life of so many living entities. LabVIEW technology made it possible to use IoT directly with an interactive interface for developers and users. The costs of implementation are very affordable since the sensors and microcontrollers are readily available at a reduced rate. The use of database technology has improved flexibility by updating all parameters continuously over a common server. The monitoring and controlling of the different parameters through the web browser and android app offer a high degree of resilience. The use of a surveillance camera provides the live video of the critical site. Thus, the proposed approach enables the remote monitoring of the plant from anywhere, and the automated control actions can be generated effortlessly by an authorized operator.

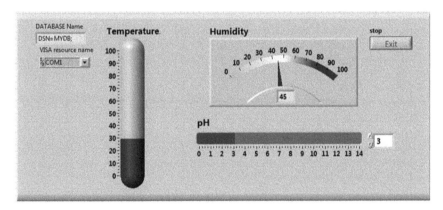

FIGURE 8.11   LabVIEW front panel.

KEYWORDS

- **Internet of Things**
- **LabVIEW**
- **XAMP**
- **Arduino**

REFERENCES

1. Halvorsen, H., Jonsaas, A., Mylvaganam, S., Timmerberg, J., & Thiriet, J. M., (2017). Case studies in IoT-smart-home solutions: Pedagogical perspective with industrial applications and some latest developments. *EAEEIE Annual Conference (EAEEIE)* (pp. 1–8). Grenoble, France.
2. Jabbar, W. A., Alsibai, M. H., Amran, N. S. S., & Mahayadin, S. K., (2018). Design and implementation of IoT-based automation system for smart home. In: *2018 International Symposium on Networks, Computers, and Communications (ISNCC)* (pp. 1–6). Rome.
3. Poongothai, M., Subramanian, P. M., & Rajeswari, A., (2018). Design and implementation of IoT based smart laboratory. In: *2018 5ᵗʰ International Conference on Industrial Engineering and Applications (ICIEA)* (pp. 169–173). Singapore.
4. Cloete, N. A., Malekian, R., & Nair, L., (2016). Design of smart sensors for real-time water quality monitoring. *EEE Access, 4*, pp. 3975–3990.
5. Roy, Das, P., & Das, R., (2018). Temperature and humidity monitoring system for storage rooms of industries. *International Conference on Computing and Communication Technologies for Smart Nation (IC3TSN)* (pp. 99–103). Gurgaon.
6. Swain, K. B., Santamanyu, G., & Senapati, A. R., (2018). Smart industry pollution monitoring and controlling using LabVIEW based IoT. *Third International Conference on Sensing, Signal Processing and Security (ICSSS)* (pp. 74–78). Chennai.
7. Al-Fuqaha, A., Khreishah, A., Guizani, M., Rayes, A., & Mohammadi, M., (2015). Toward better horizontal integration among IoT services. *IEEE Commun. Mag., 53*(9), 72–79.
8. Collina, M., Corazza, G. E., & Vanelli-Coralli, A., (2012). Introducing the QEST broker: Scaling the IoT by bridging MQTT and REST. *IEEE Int. Symp. Pers. Indoor Mob. Radio Commun.*, 36–41.
9. Al-Fuqaha, A., Guizani, M., Mohammadi, M., Aledhari, M., & Ayyash, M., (2015). Internet of Things: A survey on enabling technologies, protocols, and applications. *IEEE Commun. Surv. Tutorials, 17*(4), 2347–2376.
10. Jia, X., Feng, Q., Fan, T., & Lei, Q., (2012). RFID technology and its applications in the Internet of Things (IoT). *Int. Conf. Consum. Electron. Commun. Networks, CECNet 2012-Proc.* (pp. 1282–1285).
11. Yun, M., & Yuxin, B., (2010). Research on the architecture and key technology of Internet of Things (IoT) applied on smart grid. *Int. Conf. Adv. Energy Eng. ICAEE 2010,* (pp. 69–72).
12. Chen, B., Wan, J., Shu, L., Li, P., Mukherjee, M., & Yin, B., (2017). Smart factory of industry 4.0: Key technologies, application case, and challenges. *IEEE Access, 6*, 6505–6519.
13. Qu, T., Lei, S. P., Wang, Z. Z., Nie, D. X., Chen, X., & Huang, G. Q., (2016). IoT-based real-time production logistics synchronization system under smart cloud manufacturing. *Int. J. Adv. Manuf. Technol., 84*(1–4), 147–164.
14. Mahmoud, R., Yousuf, T., Aloul, F., & Zualkernan, I., (2016). Internet of Things (IoT) security: Current status, challenges, and prospective measures. In: *10ᵗʰ Int. Conf. Internet Technol. Secur. Trans. ICITST 2015,* (pp. 336–341).

15. Wu, Q., et al., (2018). Intelligent smoke alarm system with wireless sensor network using ZigBee. *Wirel. Commun. Mob. Comput., 2018*(2).

16. Sisinni, E., Saifullah, A., Han, S., Jennehag, U., & Gidlund, M., (2018). Industrial Internet of Things: Challenges, opportunities, and directions. *IEEE Trans. Ind. Informatics, 14*(11), 4724–4734.

17. Åkerberg, J., Gidlund, M., & Bjorkman, M., (2011). Future research challenges in wireless sensor and actuator networks targeting industrial automation. *IEEE Int. Conf. Ind. Informatics*, 410–415.

18. Bloom, G., Alsulami, B., Nwafor, E., & Bertolotti, I. C., (2018). Design patterns for the industrial Internet of Things. *IEEE Int. Work. Fact. Commun. Syst.-Proceedings, WFCS, 2018*, 1–10.

19. Nagpal, C., Upadhyay, P. K., Hussain, S. S., Bimal, A. C., & Jain, S., (2019). IIoT based smart factory 4.0 over the cloud. *Proc. 2019 Int. Conf. Comput. Intell. Knowl. Econ. ICCIKE*, 668–673.

20. Tragos, E. Z., et al., (2014). Enabling reliable and secure IoT-based smart city applications. *IEEE Int. Conf. Pervasive Comput. Commun. Work. PERCOM Work*, 111–116.

21. Jabbar, W. A., Alsibai, M. H., Amran, N. S. S., & Mahayadin, S. K., (2018). Automation system for smart home. *Int. Symp. Networks, Comput. Commun.*, 1–6.

22. Narita, H., Can, Z. Y., Mizusaki, J., & Tagawa, H., ((1995).). Solid-state CO_2 sensor using an electrolyte in the system Li_2CO_3-Li_3PO_4-Al_2O_3. *Solid State Ionics, 79*, 349–353.

23. Smoke Detector. https://en.wikipedia.org/wiki/Smoke_detector (accessed on 16 November 2021).

24. Flame Detector. https://en.wikipedia.org/wiki/Flame_detector (accessed on 16 November 2021).

25. Mesh Network. https://techiesms.com/iot-projects/nodemcu-projects/home-automation-using-mesh-networking-no-internet-no-router/ (accessed on 16 November 2021).

CHAPTER 9

Wireless Underground Sensor Networks

TULIKA GARG,[1] MANISHA BHARTI,[1] and TANVIKA GARG[2]

[1]*Punjab Engineering College, Chandigarh, India,*
E-mail: tulika.garg.bppc@gmail.com (T. Garg)

[2]*National Institute of Technology Delhi, Delhi, India*

9.1 INTRODUCTION

Sensor networks form a wide research area. They have various applications such as geology, navigation, agriculture, and security. These applications have aroused attention towards their potential to monitor underground conditions. Monitoring soil conditions, for example, mineral and soil content [1] is one of the applications of underground sensors. Underground sensors are also used to monitor landslide and earthquake (Figure 9.1) [2, 18].

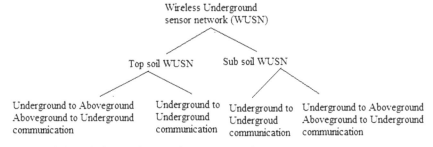

FIGURE 9.1   Wireless underground sensor network.

Harnessing the Internet of Things (IoT) for Hyper-Connected Smart World.
Indu Bala and Kiran Ahuja (Eds.)
© 2023 Apple Academic Press, Inc. Co-published with CRC Press (Taylor & Francis)

Wireless underground sensor network (WUSN) is a part of an underground sensor network in which almost all sensors' devices are deployed under the ground. There are various benefits of WUSN in comparison to conventional underground sensor networks. Some of them are explained below:

- Conventional underground sensor networks have some of their parts above the ground which can be destroyed by landscaping and agricultural equipment, for example, tractors and lawnmowers, while WUSN have all their parts under the ground, thus remaining safe from getting damaged.
- Sensor readings are often stored by data loggers for future use. WUSN helps them by wirelessly sending sensor readings to the sink.
- In conventional sensor networks tens of sensors are connected to a single data logger. Thus, if a data logger fails, it results in the failure of the entire sensor network. In WUSNs each sensor can send its reading independently and removes the need of data logger. Furthermore, WUSNs can selfheal themselves.
- In conventional underground sensor networks majority of the sensors are deployed near the data logger in order to minimize distance between them. This leads to uneven coverage density while in WUSN all sensors are deployed independently and in an even manner.

9.2   APPLICATIONS

Applications of WUSNs can be classified into four major categories. They are infrastructure monitoring, environmental monitoring, border patrol and security monitoring and location determination.

9.2.1   INFRASTRUCTURE MONITORING

Pipes, liquid storage tanks and electrical wiring are some of the underground infrastructure which needs to be monitored. WUSNs help in monitoring them. For example, fuel stations have underground tanks which contain fuel in them. These tanks need to be monitored to see that there is no leak and to check the amount of fuel present in them.

Homes which do not have sewers have septic tanks below the ground. These tanks need to be monitored to restrict the overflow. WUSNs are used in sensors-enabled underground plumbing in which sensors are deployed along the pipe's length so that they can be easily monitored and fixed if there is any leak.

Sensors can be also be used to monitor the underground parts of the bridge, building, and dams [2]. These are deployed to check stress, strain, etc. [3, 22, 28]. WUSNs can also be used for minefield monitoring. They can also be used for minefield monitoring. They can selfheal the minefield [4].

9.2.2 ENVIRONMENTAL MONITORING

WUSNs are used to check underground conditions, for example, mineral and water content of the soil. It also provides data for proper fertilization and irrigation. WUSNs are beneficial than their wireless counterparts as they can be used to give local detailed data of the soil. Sensors are also deployed inside the pots of the plants in the greenhouse setting.

WUSNs are better than the conventional agricultural WSNs. The dataloggers or surface WSN devices are not preferred in sports field monitoring while WUSNs are preferred because they can monitor soil conditions at soccer fields, grass tennis courts, golf courses and baseball fields. In all these sports, poor turf is unfavorable so maintenance of soil is important which can be done by WUSNs. Additionally, the WUSNs have underground sensors so they are not destroyed by tractors and lawnmowers.

Another important application of WUSNs is that they can monitor the presence of toxic substances. This is particularly essential for soil near aquifers and rivers which are contaminated by chemical substances discharged by industries and other sources. Thus, here underwater, and underground sensors are used to monitor the concentration of chemical substances.

Additionally, WUSNs can also be used for the landslide prediction by soil movement monitoring. The methods which are currently used are time consuming and costly. WUSNs are not expensive and their deployment is easy. WUSNs permits the dense deployment of sensors so that prediction of landslide is done at the appropriate time and people of that area can be protected.

Air quality monitoring in coal mines is also an application of WUSNs. Production of carbon monoxide and methane is a big problem which can lead to fire or explosion inside the mine, so their presence should be monitored continuously. For this hybrid architecture of underground embedded sensors and open-air sensors are deployed between the mine tunnel roof and surface of the ground. This would lead to routing of data from the mines to stations present on the surface vertically instead of taking the route through the mine tunnels (Figure 9.2).

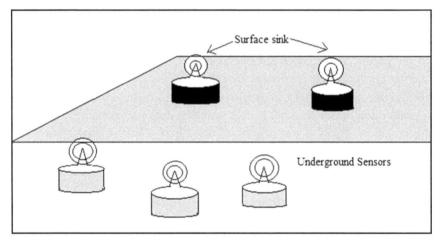

FIGURE 9.2   Deployment of underground sensors.

Monitoring and prediction of earthquake is another important application of WUSN technology. In this case, data about earthquake are attained from below the ground through multiple depths. The WUSN having multi-hop nature are deployed and they provide data to the sink which is present above the ground.

9.2.3   *LOCATION DETERMINATION OF OBJECTS*

The WUSNs can be used for determining the location of objects. One such application is that the sensors are deployed below the ground and it can communicate with the car that passes that place. It can alert the driver of the car of the traffic signal or stop sign that is upcoming. The car receives the information from the underground sensors and warns the drivers.

WUSN technology can also be utilized to find those who are buried inside the collapsed building.

9.2.4   SECURITY MONITORING AND BORDER PATROL

WUSN technology can also be utilized for the detection of the presence of people and objects above the ground. Presence monitoring uses the sensors, for example, pressure, magnetic, and acoustic sensors for this detection. This is quite helpful for checking the presence of intruders in homes. As sensors are present below the ground, intruders will not be able to know their presence and thus will not be able to disable the security system.

WUSN are also useful for border patrol. To warn the authorities of illegal crossing through the border, pressure sensors are deployed along the border length. They can give the exact location of the illegal crossing.

9.3   WUSN DESIGN CHALLENGES

WUSN design challenges include conservation of power, designing of antenna, designing of topology and extremities of environment.

9.3.1   CONSERVATION OF POWER

In order to be cost-efficient, WUSN should last for several years. In addition to this, underground channel having losses needs that these devices should have greater transmission power radios in comparison to conventional WSN devices. Thus, power conservation is a major design concern of WUSNs.

The WUSNs' lifetime is less because of the self-sufficient source of power of all devices. Furthermore, retrieving of WUSN devices is difficult thus recharging and replacement of power supply is not easy. Replacement of failed devices with new ones is also very difficult. In addition to this, conventional WSN devices can replace the traditional power supply with solar cells [5, 6, 19, 20, 25], which is not the case with WUSN devices.

In order to increase the life of the WUSN devices, it is provided with a stored power source which is quite large, but it is disadvantageous because it results in the increase of size and cost of sensors. Conservation of power can be attained by using communication protocols and hardware that are power efficient.

9.3.2   DESIGNING OF TOPOLOGY

To get the advantage of network reliability and conservation of power, the design of wireless underground sensor network topology is of utmost importance. WUSN topologies are quite different from the terrestrial WSN topologies. The application of WUSN, minimization of power usage and cost of deployment plays a crucial role in the design of WUSN topology. Thus, there should be a balance among them to have an ideal topology. The detailed discussion of each consideration is done below:

1.   **Intended Application:** Sensors of WUSN should be deployed in proximity to the phenomenon that is required to be sensed. Density of sensors at a place depends on the intended application. For example, in security applications, pressure sensors are needed to be deployed densely, while in soil monitoring, fewer devices are needed to be deployed in a particular area.

2.   **Power Usage Minimization:** Power can be conserved by designing an intelligent topology of WUSN. As path loss increases as the distance between transmitter and receiver increases, the minimization of power used can be achieved by designing the topology which has a large quantity of small hops instead of small quantity of large hops.

3.   **Cost:** Deeper the deployment of sensors more is the cost. In addition to this, extra cost will be there when the power of the sensors is finished and they are needed to be taken out for recharging and replacement. Thus, when the factor of cost has to be considered deep deployment of sensors should not be encouraged, and the quantity of sensors deployed should be minimized.

While keeping in mind all these considerations, two WUSN topologies are suggested. They are hybrid topology and underground topology:

1.   **Underground Topology:** In this topology, each sensor device is placed below the ground while sink can be placed anywhere, i.e., either above or below the ground. Underground topology can be of two types: single depth and multiple depth. In the single depth

topology, all sensors are present at the same depth while in a multi-depth topology the sensors are present at different depths. The depth of the sensor devices depends on the application for which they are used for example, soil water sensors are present deep near the plants' roots while pressure sensors are present close to the surface. This underground topology prevents the deployment of equipment above the ground thus encouraging concealment of the WUSNs.

2. **Hybrid Topology:** In this topology, sensor devices are present both above and below the ground. As wireless signals can propagate with low loss in the air than in the soil, the sensors present above the ground require lower power to send signals through a given distance than the sensors present below the ground. In addition to this, the sensors present above the ground can be easily replaced and recharged when their power gets exhausted while it is quite difficult for sensors present below the ground. The hybrid topology has a disadvantage, i.e., the full concealment of the network cannot be achieved.

9.3.3 DESIGNING OF ANTENNA

The use of appropriate antennas for devices used in WUSN is also a major challenge. The challenges included are:

1. **Variable Requirements:** Each WUSN device has different purpose thus they require different kinds of antennas. For instance, WUSN devices present in proximity to the ground may require antennas which can handle EM waves' reflection that takes place at the interface of soil and air. In addition to this, devices present near the surface can work as relays between devices present at the surface and devices present deep into the ground. Devices present deeper in the soil which can act as vertical relays need antennas that can focus both in the vertical and horizontal direction.

2. **Size:** It is known that the lower the frequency used, the larger the size of antennas is required [17]. For example, at 100 MHz frequency, 0.75 m quarter wavelength antenna is used. Thus, it is clearly shown that the size of antenna for WUSN devices is a major challenge as the sensors should be of small size.

3. **Directionality:** There is research going on to address whether a group of directional antennas or an omnidirectional antenna are

suitable for WUSN devices. Usage of Omni-directional antennas in WUSN topologies is not beneficial because WUSN devices are present at different depths and an Omni-directional antennas can create nulls at each end. This drawback can be removed by using the antennas which are oriented both in vertical and horizontal directions.

9.3.4   EXTREMITIES OF ENVIRONMENT

The underground environment is not an ideal location for WUSN devices. Insects, water, temperature extremes, and animals are not favorable for WUSN devices. Thus, there should be protection of these devices from these factors. In addition to this, the miniaturization of the WUSN devices should be done to avoid large expense and time required to deploy larger devices. Furthermore, while choosing the battery technique, the temperature of the deployment environment should be kept in mind. There is also a pressure from people and objects moving above the ground. The environmental factors that pose a challenge on the hardware deployment underground also leads to extreme underground wireless channel conditions.

9.4   UNDERGROUND WIRELESS CHANNEL

This also poses a challenge to the realization of WUSNs. Although it is appeared that underground digital communication is unexplored, EM radiation through rock and soil has been extensively researched in the past. A detailed discussion of various aspects of this wireless channel is done in this section.

9.4.1   UNDERGROUND CHANNEL PROPERTIES

Five main factors have been identified which affect communication of EM waves present underground. They are extreme path loss, multi-path fading, noise, reflection/refraction, and reduced propagation velocity:

1. **Extreme Path Loss:** Material absorption is a major factor of path loss when EM waves are used for underground communication. Losses are also due to the frequency of the EM waves and the soil

or rock properties through which EM waves travel. Lower frequencies lead to less attenuation in comparison to higher frequencies. Path losses are also due to water content present in the soil and its types. Soil is divided into different classes based on their particles' size. In decreasing order, they are: sand, silt, and clay [7]. EM waves can travel easily through the sandy soils than through the clay soils. Furthermore, the increase in the content of water in the soil also produces a large amount of losses.

2. **Refraction/Reflection:** WUSN devices which are placed close to the surface can communicate with devices present at the surface and underground, for example, a sink present at the surface. Thus, it is necessary to have a communication link which is present partially in the air and partially underground. The traveling EM waves when reaches the air-ground surfaces, it is partially passed through the surface and partially reflected back.

3. **Multi-Path Fading:** The reflection and refraction of the EM waves at medium transitions also result in multi-path fading. The plants' roots, rocks present underground and soil properties having varying nature can also scatter the EM waves and thus leads to multi-path fading.

4. **Reduced Propagation Velocity:** The velocity of the EM waves traveling through the soil or rock (which act as dielectric material) gets reduced as compared to that in the air. The range of dielectric constant of soils are from 1 to 80 and EM waves can experience the reduction of velocity as low as 10% of the speed of light.

5. **Noise:** Underground channel is also affected by the noise. Underground noise is present in the form of lightening, power lines and electric motor [8, 21, 26, 27]. Noise present under the ground has low frequencies (below 1 kHz).

9.4.2 EFFECT OF SOIL PROPERTIES ON THE UNDERGROUND CHANNEL

The complex dielectric constant is formed by the composition of soil which includes particle sizes, temperature, water content and density. The dielectric constant leads to the attenuation of electromagnetic wave traveling through the soil and it is necessary to know its value. The effect of these parameters on attenuation of the signal is discussed in great detail below:

1. **Water Content:** This of soil is an important parameter which has to be considered when the signal loss through the soil has to be predicted. The channel becomes more lossy when the water content in the soil increases. The type of soil also decides the effect of water content on the losses, for example, clay soil shows more attenuation when water content is increased as compared to sandy soil. Attenuation also depends on the frequency that is being used. At a given water content, the frequencies with higher value experience more attenuation in comparison to the frequencies with lower value.
2. **Particle Size:** Diameter of soil particles are used to classify the soil. Sandy soil provides more attenuation to the signal as compared to silt or clay soil [9, 23].
3. **Density:** Path loss increases as the density of soil increases.
4. **Temperature:** Temperature of soil alters the properties of dielectric and will result in increased signal attenuation [7].

9.4.3   *SOIL DIELECTRIC PREDICTION MODELS*

Using EM relations, the attenuation of the traveling EM waves through the soil or rock can be predicted by knowing the dielectric constant of that soil or rock. This does not give an accurate model, but it can provide a good intuition of the conditions of channel. However, having the knowledge of the dielectric constant of the soil where WUSN devices are placed is still a major challenge. Furthermore, some models are able to accurately predict the dielectric constant for a homogeneous soil sample but path loss prediction for the underground channel still remains a big challenge because in reality soil are inhomogeneous in nature. Density, soil make up and water content can vary very quickly [9].

Soil dielectric prediction models are divided into three classes: volumetric, phenomenological, and semi-empirical. Relation of the frequency-dependent behavior of soil with relaxation times is done in phenomenological models. Volumetric models compute dielectric constant by knowing the dielectric properties of each material and the makeup of soil. Semi-empirical models compute dielectric constant by observing the relationship between material dielectric properties and its characteristics (Table 9.1).

TABLE 9.1   Soil Properties and Their Effect on Signal Attenuation

Parameter	Change	Effect on Signal Attenuation
Water content	Increases	Increases
Temperature	Increases	Increases
Soil bulk density	Increases	Increases
% Sand	Increases	Decreases
% Clay	Increases	Increases

9.4.4   ALTERNATIVE PHYSICAL LAYER TECHNOLOGIES

From the above discussion, it can be inferred that underground is not suitable for wireless communication. Particles of soil and water content present inside the ground causes high attenuation of EM waves. Although EM waves are chosen due to their capability to carry a large amount of information below the ground, other possible unexplored physical layer options are present for WUSNs.

First possible option for underground communication is Magnetic Induction (MI). There are various benefits of using MI. Soil and water causes very little attenuation of magnetic field as to EM waves. Although MI is not preferred for communication in open-air as the strength of magnetic field falls as $1/R^3$ where R indicates the distance from the transmitter, the reduction in attenuation of signal compensates for this in the underground.

In addition to this, MI does not use antennas rather it uses a small coil of wire for the transmission and reception of signals. The magnetic field strength depends on three factors. They are: number of wire turns, the material's magnetic permeability present in the coil's core and cross-sectional area. The use of wire coils in MI has various benefits over the antennas used for electromagnetic waves propagation. The low frequency of EM waves makes the size of antennas very large which is a major drawback as small underground sensors are preferred.

9.4.5   EXISTING WORK

In the past, the main focus was on the research of single-hop underground communication links but it died out because long-distance links were not feasible [10, 11]. For instance, a system is introduced in Ref. [11], where people who are trapped in mines could communicate by the use of electrodes. The same setup is used by the receiver. The frequencies from 1 to 10 kHz are used by the system. The data rate of 2 kbps is achieved by the system. In Ref. [12], it has been found that TWSN motes are feasible. The Crossbow's MicaZ motes which uses a frequency of 2.4 GHz and carries out the transmission at 1 mW of power are buried deep inside the ground. The work related to the usage of FSK modulation in MI has been demonstrated in Ref. [13].

9.5 COMMUNICATION ARCHITECTURE

The protocol stack of WUSNs is explained in this section. Figure 9.3 shows the protocol stack, its five layers and cross-layered task management and power management.

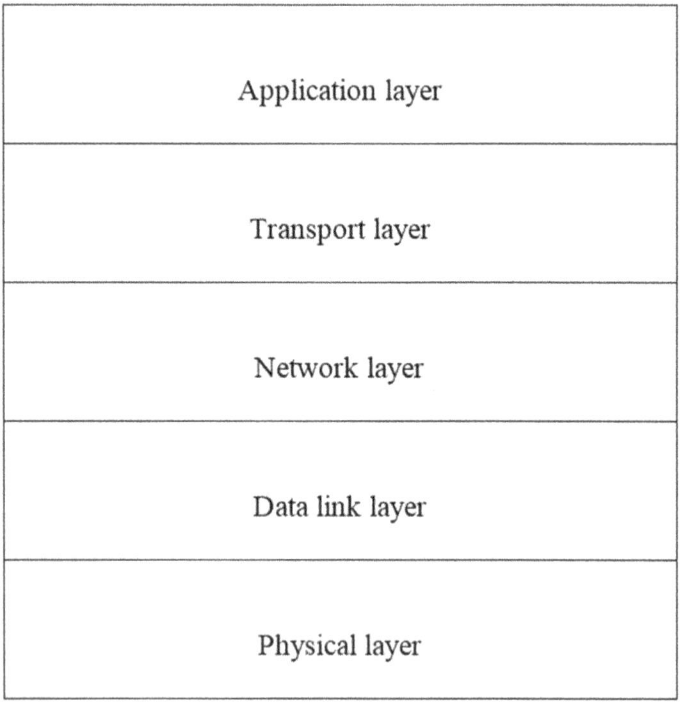

Application layer
Transport layer
Network layer
Data link layer
Physical layer

FIGURE 9.3 Protocol stack of WUSN.

9.5.1 PHYSICAL LAYER

Physical layer communication poses a major challenge for wireless underground sensor networks. As discussed before, EM signals traveling through rock and soil are greatly attenuated. Additionally, the underground environment changes with time which poses another challenge for example, path loss depends on the soil properties such as water content which varies with time. Thus, attenuation is caused by wet soils to the extent of making communication impossible through them [14, 24]. After the rainfall, water content in the soil increases and thus path loss is incurred by it which can last for a large amount of time. Because of the challenges discussed above, it is very important to design an efficient antenna. Placing an antenna in the soil which is a conductive medium can affect its characteristics of reception and radiation [15]. As discussed before, EM waves with lower frequencies are less attenuated as compared to higher frequencies when they propagate through soil or rock [16], so EM propagation is only feasible in the lower frequencies.

As specified that there are power constraints in WUSN and lower frequencies are used to reduce the losses, the use of suitable modulation scheme in WUSN is another major challenge. Basically, analog modulation is used in underground communication but [8] shows that digital modulation schemes such as QAM-16, QAM-32 and QPSK are also used. Apart from this modulation schemes for underground communications are still unexplored. Research challenges at the physical layer are:

- In order to make out the proper physical layer technology an analysis is carried out of MI (magnetic induction), electromagnetic, and seismic communication in the underground. The various combinations of such technologies might be ideal, specifically for shallow sensor devices which carries out the communication with both surface sensors and those present below the ground.
- A modulation scheme which is power efficient for the dynamic underground high-loss underground channel must be selected. There is a need for the research in various modulation schemes which depends on underground channel conditions. The channel gets impaired after a rainfall when the channel is severely impaired, for instance, the situation gets better if higher data rates trade of for an easier modulation scheme.

- The examination for the trade-off between capacity and reliability must be carried out. For a particular underground distance, the propagation of lower frequencies with lower loss must be carried out, but they also have lesser bandwidth available for transmission of data which reduces the capacity of the channel.
- For the underground wireless communication channels' capacity, a theoretical study is required.

9.5.2   DATA LINK LAYER

The MAC protocols which are used in terrestrial WSNs will not perform efficiently in wireless underground sensor networks. The protocols which are used in terrestrial wireless sensor networks are either based on TDMA or based on contention and their aim is to minimize consumption of energy by addressing four primary areas: overhearing, idle-listening, collisions, and control packet overhead [17]. WUSNs require special kind of MAC protocol because of the underground characteristics which we have discussed before.

As consumption of energy is the aim of MAC protocols used for terrestrial WSNs, saving of energy can be done by decreasing the time of idle listening [14]. In WUSNs, antennas must transmit EM waves which have much higher power so that they can overcome the path losses. In addition to this, for acceptable lifetime WUSNs should minimize the number of transmissions.

WUSNs MAC protocol should reduce collisions as these collisions lead to retransmissions. This can be achieved by using contention-based protocol along with RTS/CTS scheme. TDMA-based scheme can also be used to reduce collisions. This can be achieved by allocating a time slot for each device. In this case synchronization is the problem which has its own overhead.

Research challenges at the data link layer are:

- In order to determine which of the protocols, whether contention-based or TDMA-based MAC is more suitable for wireless underground sensor networks, one has to realize the tradeoffs between the energy savings and the additional overhead of a TDMA-based protocol. A hybrid MAC scheme is a possible solution to this.
- One needs to explore the synchronization (in minutes).
- One needs to explore the adaptive FEC schemes to find out whether it is a possible solution for the underground channel's unique nature.

More efficient FEC schemes are required when the damages are caused by wet soil, in order to overcome the losses.

- The determination of the optimal packet size is needed to be done. This should be done particularly taking into consideration the underground channel effects in the calculations. There is an important role of the packet size in both quality of service and power conservation. The power efficiency should be maximized by minimizing the amount of overhead sent as packet headers. This will result in the larger packet size but the outcome of the thesis is the increase in the overall latency in the larger packet sizes. The reason is that the time for which it has to wait for the channel will become longer. So, it is required to find out an ideal tradeoff among all these factors.

9.5.3   NETWORK LAYER

Ad-hoc network routing protocols have three categories: proactive, reactive, and geographical. In proactive routing protocols, routes are continuously maintained among all the devices present in the network, while in reactive protocols, routing is established only when it is required. Both these categories have significant signaling overhead.

In geographical routing protocols, routing is established by knowing the device physical location. Thus, in this way route can bring data nearer to the destination with each hop. Most of the sensors are deployed by digging a hole for each one of them. Therefore, an exact location of the sensor can be recorded at the deployment time. In this scenario, geographical routing protocols are preferred.

Research challenges of the network layer include:

- It is required to examine how the low duty cycle of WUSNs is affecting the routing protocols. There could be drastic change in the network topology; hence the network layer and sensing intervals should handle this efficiently.
- There is a need for the research on routing protocols which is suitable for time-sensitive WUSN applications, for example, presence monitoring for security with underground pressure sensors.
- It is required to examine whether the multi-path routing algorithms are applicable or not. If there is a link failure, then these algorithms

will not use the complete path switching. The research is required to ensure that they are energy-efficient.

9.5.4   TRANSPORT LAYER

WUSNs need a transport layer to attain reliable collective transport and to conduct congestion control and flow control. Various transport layer protocols are developed for terrestrial wireless sensor networks, but they need to be re-examined for the underground channel because of high loss rates.

The main aim is to protect limited sensor resources and to increase the efficiency of networks. Therefore, to protect the network from congestion of data, control of congestion is required and to not to use network devices with excessive data, flow control is needed. As WUSNs have low data rate, congestion is a big problem especially near sink. To avoid this, we can route the data to terrestrial relays which have high data rate. All these things can be accomplished at the network layer. Various TCP implementation are not suited for WUSNs because functionality of flow control is based on a window-based mechanism and retransmissions. As it is already stated before retransmissions are not suited in WUSNs because they need more energy. Rate based transport protocols are also not suited for WUSNs.

Because of these reasons, it is quite essential to have new strategies to attain reliability and flow control under the ground.

Research challenges at the transport layer are:

- For the sensor reporting rate, novel optimal policies are required to avoid congestion and to increase the transport reliability in WUSNs;
- Novel transport reliability definitions need to be developed on the underground channel model;
- In WUSNs, an acceptable loss rate needs to be specified. This leads to power savings by decreasing the re-transmissions which is essential for these power constrained WUSN devices.

9.6   CONCLUSION

This chapter discussed about WUSNs where all devices are deployed below the ground. WUSNs advantages over conventional underground sensors are discussed. Various applications of WUSNs are also discussed such as monitoring of sports field and garden, border monitoring, minefield monitoring, plumbing, and security. Various design challenges of WUSNs are discussed power conservation, topology design, antenna design. Aspects of underground wireless channel are discussed in detail and the chapter is concluded by discussing challenges related to communication protocol stack of WUSNs.

KEYWORDS

- **underground sensors**
- **conventional sensor network**
- **protocol stack**

REFERENCES

1. Advanced Aeration Systems, Inc. Rz-Aer tech sheet. Available from: http://www.advancedaer.com/ (accessed on 16 November 2021).
2. Park, C., Qiang, X., Pai, H. C., & Masanobu, S., (2005). Duranode: Wireless networked sensor for structural health monitoring. *Sensors*, 4. IEEE.
3. Cheekiralla, & Sivaram, M. S. L., (2004). *Development of a Wireless Sensor Unit for Tunnel Monitoring*. PhD diss., Massachusetts Institute of Technology.
4. Rolader, G. E., John, R., & Jad, B., (2004). Self-healing minefield. *Battlespace Digitization and Network-Centric Systems IV (International Society for Optics and Photonics), 5441*, 13–24.
5. Jiang, X., Joseph, P., & David, C., (2005). Perpetual environmentally powered sensor networks. *IPSN, Fourth International Symposium on Information Processing in Sensor Networks IEEE*, 463–468.
6. Voigt, T., Hartmut, R., & Jochen, S., (2003). Utilizing solar power in wireless sensor networks. In: *28th Annual IEEE International Conference on Local Computer Networks* (pp. 416–422).
7. Daniels, D. J., (1996). Surface-penetrating radar. *Electronics & Communication Engineering Journal, 8*(4), 165–182.
8. Vasquez, J., Victor, R., & David, R., (2004). Underground wireless communications using high-temperature superconducting receivers. *IEEE Transactions on Applied Superconductivity, 14*(1), 46–53.
9. Miller, T. W., Brian, B., Jan, M. H. H., Hong, S., Louis W. D., & Coen, J. R., (2002). Effects of soil physical properties on GPR for landmine detection. *Fifth International Symposium on Technology and the Mine Problem, 1*–10.

10. King, R. W. P., & Charles, W. H. Jr., (1968). The transmission of electromagnetic waves and pulses into the earth. *Journal of Applied Physics, 39*(9), 4444–4452.

11. Williams, H. P., (1951). Subterranean communication by electric waves. *Journal of the British Institution of Radio Engineers, 11*(3), 101–111.

12. Stuntebeck, E., Pompili, D., & Melodia, T., (2006). *Underground Wireless Sensor Networks Using Commodity Terrestrial Motes, Poster Presentation (IEEE SECON).*

13. Sojdehei, J. J., Paul, N. W., & Donald, F. D., (2001). Magneto-inductive (MI) communications. *An Ocean Odyssey, Conference Proceedings IEEE, 1,* 513–519.

14. Bala, I., Bhamrah, M. S., & Singh, G., (2017). Capacity in fading environment based on soft sensing information under spectrum sharing constraints. *Wireless Networks, 23,* 519–531. Springer.

15. King, R. W. P., Glenn, S. S., Margaret, O., & Tai, T. W., (1981). *Antennas in Matter: Fundamentals, Theory, and Applications.* 29690. STIA 81.

16. Wait, J., & Fuller, J., (1971). On radio propagation through earth. *IEEE Transactions on Antennas and Propagation, 19*(6), 796–798.

17. Kredo, II. K., & Prasant, M., (2007). Medium access control in wireless sensor networks. *Computer Networks, 51*(4), 961–994.

18. Bala, I., Bhamrah, M. S., & Singh, G., (2017). Rate and power optimization under received-power constraints for opportunistic spectrum-sharing communication. *Wireless Personal Communications, 96,* 5667–5685. Springer.

19. Roundy, S., Dan, S., Luc, F., Paul, W., & Jan, R., (2004). Power sources for wireless sensor networks. *European Workshop on Wireless Sensor Networks, 1*–17.

20. Akyildiz, I. F., Weilian, S., Yogesh, S., & Erdal, C., (2002). Wireless sensor networks: A survey. *Computer Networks, 38*(4), 393–422.

21. Zhang, Y. P., Guo, X. Z., & Sheng, J. H., (2001). Radio propagation at 900 MHz in underground coal mines. *IEEE Transactions on Antennas and Propagation, 49*(5), 757–762.

22. Lynch, J. P., (2002). *Decentralization of Wireless Monitoring and Control Technologies for Smart Civil Structures.* PhD diss., Stanford University.

23. Jaramillo, D. F., Dekker, L. W., Ritsema, C. J., & Hendrickx, J. M. H., (2000). Occurrence of soil water repellency in arid and humid climates. *Journal of Hydrology, 231,* 105–111.

24. Trang, A. H., (1996). Simulation of mine detection over dry soil, snow, ice, and water. *Detection and Remediation Technologies for Mines and Mine like Targets, 2765,* 430–440.

25. Kansal, A., Dunny, P., & Mani, B. S., (2004). Performance aware tasking for environmentally powered sensor networks. *Proceedings of the Joint International Conference on Measurement and Modeling of Computer Systems,* 223–234.

26. Holloway, C. L., David, A. H., Roger, A. D., & George, A. H., (2000). Radio wave propagation characteristics in lossy circular waveguides such as tunnels, mine shafts, and boreholes. *IEEE Transactions on Antennas and Propagation, 48*(9), 1354–1366.

27. Bala, I., Bhamrah, M. S., & Singh, G., (2019). Investigation on outage capacity of spectrum sharing system using CSI and SSI under received power constraints. *Wireless Networks, 25,* 1047–1056. Springer.

28. Chase, S. B., & Emin, A. A., (2001). *Health Monitoring and Management of Civil Infrastructure Systems,* 4337. SPIE.

CHAPTER 10

Energy Harvesting Techniques for Future IoT Applications

N. VITHYALAKSHMI,[1] G. S. VINOTH,[2] H. D. PRAVEENA,[1] and
P. AVIRAJAMANJULA[3]

[1]ECE, Sree Vidyanikethan Engineering College, Tirupati, Andhra Pradesh, India, E-mail: vidhyavinoth@gmail.com (N. Vithyalakshmi)

[2]QUEST Global, Technopark, Thiruvananthapuram, Kerala, India

[3]Professor, EEE, PREC, Thanjavur, Tamil Nadu, India

10.1 INTRODUCTION

Nowadays, many investments are made within the Internet of Things. All electronic gadgets like personal computers and smartphones are not only connected through the internet, its connectivity with billions of things and devices. Whatever the applications, all electronic devices require some amount of power to operate. IoT mainly consists of wireless sensors to collect data. Energizing the devices is a huge challenge. Almost all low-power devices like remote sensors, embedded systems are usually powered up by batt\eries. But most of the battery has a limited lifetime, for long-lasting of devices; batteries should be changed frequently. This replacement is very costly and difficult when there are several devices in isolated location. Harvesting energy is one of the best solutions to battery for boundless functioning life cycle of low power devices, where it is costly and difficult to replace. Energy harvesting means extracting or scavenging energy from non-traditional fundamental sources like mechanical, solar, thermal, RF, and wind energy are captured and deposited for future usage.

Harnessing the Internet of Things (IoT) for Hyper-Connected Smart World.
Indu Bala and Kiran Ahuja (Eds.)

This chapter discusses the IoT growth, importance, technologies, applications, elements of energy harvesting, possible sources and quantities of energy that can be collected by using technologies like thermoelectric generators, piezoelectric elements, electromagnetic generators, photovoltaic, photodiode, etc. The deliverable energy level of different technologies varies from μw to mw range for specific device sizes and respected work. IoT expresses net of physical entities which are embedded with software. Sensors connect and exchange data with other circuits and systems through the internet. More than 7 billion devices connected to IoT ranging from household objects to modern industrial tools. The professional expecting this number will grow up to 22 billion by 2025 [1].

10.2   WHY IoT IS SO SIGNIFICANT?

Internet of Things has turned out to be one among the foremost essential technology to connect everyday applications like automotive, thermostats, appliances used in the home, infant monitors to the internet through smart devices. Faultless exchange of information is feasible among people, methods, and things. Due to less cost computation of mobile technologies, cloud, big data analytics, data can be collected and shared with minimum human involvement from physical things. In hyperlinked domain, smart systems can store, regulate, and monitor the information transfer between connected objects as shown in Figure 10.1.

10.3   TECHNOLOGIES USED IN IoT

Even though IoT has been active for an extended period, the recent technical advances have made it possible, as shown in Figure 10.2:

1. **Low Power Sensor Technology:** IoT technology is made possible for more manufacturers due to the use of reliable and reasonable sensors.
2. **Connectivity of Networks:** Effective data communication in-between sensors and cloud through internet has become easier due to latest web network protocols.
3. **IoT Platform:** Growth in the accessibility of cloud computing has allowed consumers and commercial people to use the platforms and scale them up without managing it.

4. **Expert System and Data Analysis:** From the various and huge amount of data available in the cloud and with better developments in expert systems and analytics has provided businesses easy and faster insights. The data created by IoT the advent of associated technologies has pushed the limits of IoT and these technologies too.

5. **Artificial Intelligence:** Development in Neural networks has made natural-linguistic processing to IoT systems and has made them less expensive, attractive, and feasible for using at home. Advances in neural networks have brought natural-linguistic processing to IoT devices and made them appealing, affordable, and viable for home use.

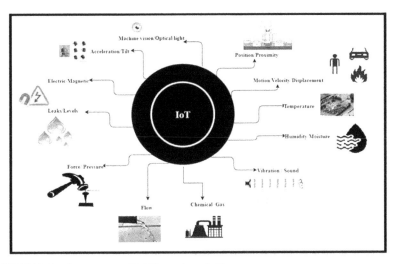

FIGURE 10.1 Importance of IoT in various fields.

10.4 GROWTH OF IoT

The rapid progress of the IoT is disturbing almost every industry. The business models of innovative firms use data these industries generate. Both service providers and device manufacturers are offered significant opportunity to afford solutions that initiate functional efficiency and revenue gain for their respective customers by hyper-connected IoT [1]. A crucial point happened in 2008 is that the human population was surpassed by the things connected to the web. The acceptance percentage of IoT is five times quicker than the acceptance of electric energy and telephony as depicted in Figure 10.3. This equates to nearby six gadgets per individual throughout globe. One of the exciting trends in the progression of IoT is its shift from IP V4 that is consumer-based to IPv6, which is operational technology specifically based on machine-to-machine communications, which includes smart objects, sensors, and combining hardware and software clusters.

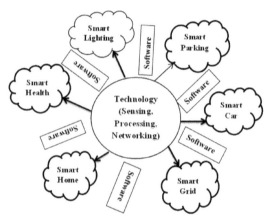

FIGURE 10.2   Internet of Things.

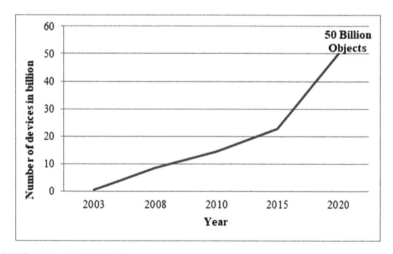

FIGURE 10.3   IoT growths.
Source: Cisco security Ref. [31].

The improvements and development in IoT technologies has allowed the growth of smart devices as tabulated in Table 10.1. As an example, a sensor which can compute, store, and has network competency, built with less energy consumption and very less form factor. Early starters and Investigators are encouraged by developments in wireless techniques that include satellites and radio, reducation of device sizes and industrial development, computing, increment in transmission capacity and storing power. All of the above provides a chance that lessens managing and functioning costs through changing the systems from legacy protocols.

TABLE 10.1   IoT Market Size in India

S. No.	Applications	IoT Market Size India
1.	Utilities	25%
2.	Manufacturing	18%
3.	Transport and logistics	13%
4.	Automotive	11%
5.	Healthcare	10%
6.	Retail	9%
7.	Agriculture	8%
8.	Others	3%

10.5   THE BENEFIT OF IoT IN INDUSTRIES

Organizations that are best suitable for IoT are the ones that will benefit from using sensors in their business developments.

10.5.1   IoT IN MANUFACTURING

Competitive advantages can be gained by manufacturers by monitoring production lines to support preventive maintenance when sensors find out an imminent failure on equipment's. If any compromise occurs in production, sensors will measure the deviations. With such sensor alerts, companies can speedily check machines for precision and replace it from manufacturing until the removed one is restored. This enables manufacturers to reduce working rates, get good uptime, and increase asset performances.

10.5.2   IoT IN AUTOMOTIVE

This industry is about to grasp the significant gains by IoT utilization. Apart from the advantages of IoT in production lines, sensors can identify breakdown in vehicles on the road and can send alertness to the driver.

10.5.3   TRANSPORTATION MANAGEMENT

The logistics and transportation industry enjoys the widespread use of IoT. Fleets of automobiles like cars, trucks, ships, and trains which carry goods and stock can be rerouted depending on climatic conditions, driver, or vehicle availability. Inventory also could be fitted with sensors to monitor and control temperature, tracking activities, etc. Eatables, drinks, flower, and drug industries that regularly transfer temperature sensitive goods could benefit significantly by IoT applications. Notifications will be provided by these applications while the temperature level changes more or less which can be a threat to the product.

10.5.4   IoT IN RETAIL TRADE

Retail businesses use IoT applications to maintain inventory, improve consumer experience, reduce functioning costs, and enhance supply chain. For instance, intelligent racks fitted with load sensors that collect and send information to the computing platform can observe inventory and generate alerts automatically if goods are running less.

10.5.5   GENERAL PUBLIC SECTOR

IoT offers a wide range of benefits within service-based surroundings and public sectors. IoT-based applications are used in Government utilities in order to alert their users in mass outages, even in smaller disruptions like water, sewer services or power. IoT applications will collect data on outages and can organize resources to assist services get over failures at greater speed.

10.5.6   IoT IN HEALTHCARE SYSTEM

Sensor-based IoT patient monitoring support system provides numerous advantages to the healthcare sector. Nurses, hospital attendants, and Doctors will get to know the precise location support systems required for patients such as wheelchairs. If supporting systems are connected with smart sensors, it can easily be tracked by sensor monitoring platforms. Hence anyone who is trying to find a support system can quickly access the closest available support system. In many hospitals, the patient support system is often tracked in this way to make sure apt usage and in order to maintain the physical count of asset in each division.

10.5.7   SAFETY IN INDUSTRIES

Apart from tracking physical assets, smart industries are showing interest to improve worker and their working environment protection. Employees who work in hazardous atmospheres like power plants, oil fields, chemical, gas fields, and mines must know the incidence of a dangerous event that will affect them. Once IoT sensors are connected to the employees, then they could be rapidly alerted and set free quickly from accidents. Also, smart sensors are used in wearable's which will detect human health conditions and nearby environmental situations. This application helps people to recognize their individual health and also allow doctors to monitor and study them remotely.

10.6   COMMON TYPES OF IoT SENSORS FOR ENERGY HARVESTING APPLICATIONS

10.6.1   TEMPERATURE SENSING ELEMENT

These sensors measure the degree of hotness produced by an object or nearby zone. Their applications are in refrigerators, air conditioning systems and gadgets used for controlling the environment. They are additionally utilized in production, health monitoring and farming. Thermal reading sensors are often used in all IoT platforms, from agriculture to production and also used to observe the temperature in soil, machines, plants, and water. Thermo sensors comprise of resistor temperature detectors, thermistors, integrated circuits, and thermocouples. Few general temperature sensors types are shown in Figure 10.4.

RTD

Thermocouple

FIGURE 10.4   Temperature sensing element.
Source: Sani Theo, Ref. [33].

10.6.2   HUMIDNESS SENSORS

Quantity of water vapor in air or moisture can cause distress to human comfort and also several manufacturing progressions in production companies [26]. Hence observing humidity range is vital. Generally used units for the measurement of humidity are dew factor, relative humidity, or frostiness and ppm. Figure 10.5 shows conventional humidness sensor.

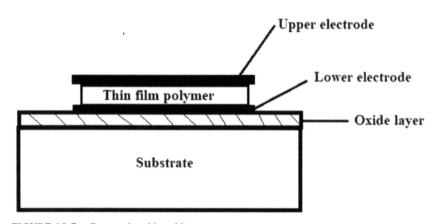

FIGURE 10.5   Conventional humidness sensor.

10.6.3 MOVEMENT SENSORS

Apart from security purposes, motion sensors are used in automated toilet flushers automated sinks, automatic door controls, energy management systems, automatic parking systems, hand dryers, etc. These sensors can be used in IoT and can be observed by using a computer or smartphone. The popular passive infrared motion sensor is shown in Figure 10.6.

FIGURE 10.6   PIR movement detection sensor.
Source: Sani Theo, Ref. [33].

10.6.4 GAS SENSING ELEMENT

They are used to identify the deadly gases emitted to nature from industries, house appliances, automotive, etc. Most commonly used sensing technologies are semiconductor, photoionization, and electrochemical. With methodological improvements and the latest specs, there are lots of gas sensors present to spread wireless and wired connections that are used in smart IoT applications shown in Figure 10.7.

10.6.5 SMOKE SENSORS

Smoke detectors are used for a quite long period of time in houses and industries. The advancement of IoT has turned the above applications very simple and more ease to use. Moreover, adding a smart wireless system to smoke detector supports added features which increase security and accessibility.

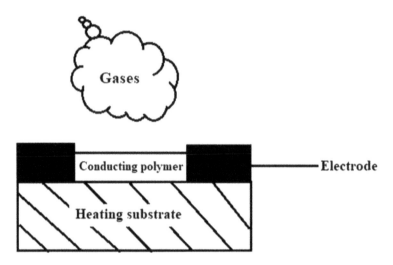

FIGURE 10.7　Gas sensor.

10.6.6　PRESSURE DETECTOR

They are utilized in smart systems to repeatedly observe and verify the systems driven by applying pressure. The device issues a warning to the user when the pressure level is above a threshold limit, which is to be fixed. As an example, BMP-180 is a standard digitized pressure sensor used in outdoor equipment, GPS navigation systems, PDAs, and mobile phones. Altitude and force in aircrafts and smart vehicle are measured by pressure sensors. Tire pressure observing system employed will warn the driver if the pressure of the tire is very low and which will cause risky driving situations. Figure 10.8 shows basic pressure sensor.

10.6.7　IMAGE SENSORS

Devices like radars, photographic equipment, media, medical diagnosis; sonars, twilight vision devices; thermographic systems and biometric systems generally use these sensors for effective functioning. Within small scale industry, sensors monitor clients visiting the shop or malls with the help of an IoT system. In commercial buildings and work places, IoT network is used to supervise the employees.

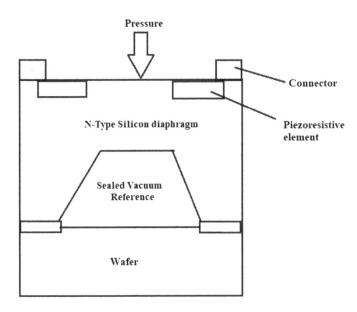

FIGURE 10.8 Pressure detectors.

10.6.8 ACCELEROMETER SENSORS

Applications like aircrafts, smartphones, vehicles, and other applications are utilized these sensors to identify the alignment of an item, motion, positioning, tap, vibration, and inclination. Nowadays, various types of accelerometers available in the market to fulfill our needs, among them piezoelectric, hall-effect, and capacitive accelerometers are popular. Figure 10.9 shows basic accelerometer sensor.

10.6.9 IR SENSORS

It is a microelectronic device that is utilized to sense definite features of nearby environment either by means of detecting or radiating infrared signals. Also, IR sensing element can measure hotness released by objects and their movement. Most of the sensor varieties measure infrared signals, instead of radiating them. Various IoT applications uses IR sensor that compromises the medical sector to observe blood pressure and flow of blood; mobile phones are used to operate as a remote controller and additional utilities and wearable's sense capacity of light (Figure 10.10).

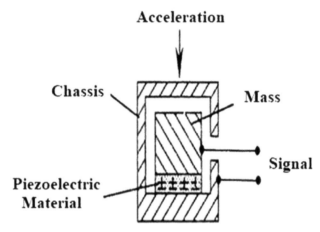

FIGURE 10.9　Basic accelerometer sensors.

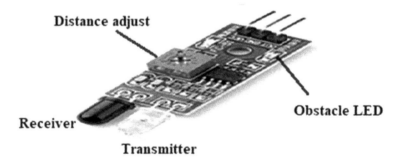

FIGURE 10.10　IR sensor.
Source: Sani Theo, Ref. [33].

10.6.10　*PROXIMITY SENSORS*

This kind of sensors normally detects the presence or absence of a neighboring object devoid of physical interaction. Proximity sensing element repeatedly radiates an electromagnetic waves or beams of electromagnetic signals to detect the variations in the field. Process flow control and observation, gadgets counters are commonly used applications. Different categories of proximity sensing elements like magnetic, inductive, photo-electric, capacitive, and ultrasonic. Capacitive proximity sensor is given in Figure 10.11.

FIGURE 10.11 Capacitive proximity sensor.
Source: Sani Theo, Ref. [33].

10.7 PROGRESSIVE SENSOR TECHNOLOGY TO IMPROVE OUR DAY-TO-DAY LIFE

To realize sensor functioning, let us find how recognizing works within the physical body. Humans experience the outside world through various sensing methods like visualizing, hearing, taste, touch, and smell. All these experiences are giving sensory information around the environments that transmits to the brain through the nervous system [1]. Now brain adopts to react to a certain situation or exposure. Nervous systems do not decide any difficult conclusions regarding data it carries and is concluded by the brain only. Based on sensory organs input, the brain directs out the data as a response to that particular input is shown in Figure 10.12. Here some examples are explained how sensor technology used in our daily life.

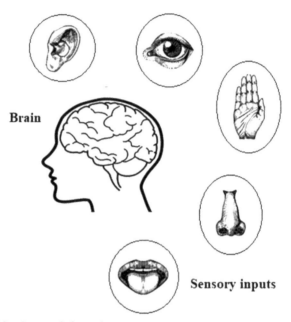

FIGURE 10.12   Sensory information.
Source: Kaivan Karimi, Ref. [1].

10.7.1   PEDOMETER TEST

Customary pedometers use pendulum, which has to be fixed at an angle in the hip to evade incorrect evaluations. For example, while walking, pedometer calculates every step by tracking the pendulum oscillating with respect to the hip movement and registering a counter every swing. But incorrect analyzes are common because of deviations in step, the climbing angle during walk.

Micro-electro mechanical system-based pedometers used accelerometers that executed 1D, 2D, or 3D axis finding person's movement, by precisely calculating footstep readings in contrast to the earlier pedometers which simply record one step every swing. Accelerometer calculates an individual's actions repeatedly every other second. Our aim is not to simply calculate the step counts, but also exactly determine the amount of energy get rid of while walking staircases or hilltops [1]. Nowadays mobile phones also have a pedometer function which could be fitted in arm. With the mixture of sensing systems like gyroscope, altimeter, and accelerometer and smart control system in order to calculate and evaluate the data, provides a very precise pedometer as shown in Figure 10.13. Exact pressure evaluations are also required for temperature calculations, and hence temperature compensation system is usually included to provide accuracy. The motion of the arm introduces parasitic movement which is measured by gyroscope.

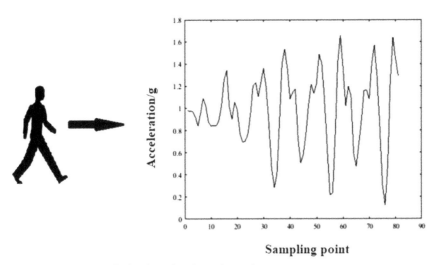

FIGURE 10.13   Analysis of acceleration using pedometer.
Source: Kaivan Karimi, Ref. [1].

10.7.2   GAMING PLATFORM

Below is an example of gaming platform that detects emotion by detecting and collecting data from different physical variations and states, like:

- Relaxation of muscle measured through pressure sensor;
- Heartbeat change analysis through ECG;
- Perspiration measured through capacitive sensing element;
- Attitude monitoring by means of an accelerometer sensor;
- Muscle contraction calculated via a pressure sensor.

Analysis of emotions within the gaming platform as shown in Figure 10.14. Data collected from various sensors utilized in the gaming platform given to the smart control devices to detect emotions and provides the player response for the duration of game situations to make the game more thrilling.

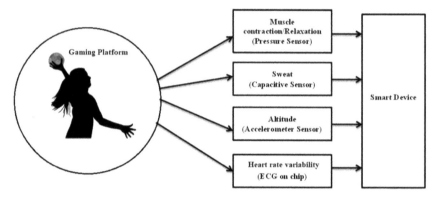

FIGURE 10.14   Emotions sensing in gaming platform.
Source: Kaivan Karimi, Ref. [1].

10.8   ENERGY HARVESTING TECHNOLOGY

Nowadays, sensors are major building blocks for IoT. Sensors get data from environment and transfer to the control unit. These sensors need power supply so as to work [2]. Sensors requires very small amount of power to work. Generally, batteries are used for several years to power these devices, but it is difficult to power the sensors using batteries. Similarly, biomedical devices [3] and the other low power applications require a bit of power for their operation. Earlier, powering of such low power devices is going to be done through batteries. These batteries got to be recharged frequently. Even long-term batteries should be recycled every few years and so to use batteries to power up low power application circuits is not an efficient method. There is an efficient method available to power such sort of low power devices called Energy Harvesting method [4].

Energy harvesting is a technique that captures and convert a small amount of energy that is available in the nearby surroundings to useful electrical power. Electricity is either directly used or collected and preserved for future use. This becomes another power source for applications where grid power is not available and is not sufficient for windmills or solar cells.

Mostly low-power systems, like remote sensing elements and embedded applications, are energized through batteries. But long-life batteries have reduced lifetime and should be changed frequently. These substitutions are costly if many sensing elements are used in remote places. Power harvesting technique provides unlimited lifetime for the low-power devices which will remove the requirement to change batteries as it is expensive, unrealistic, or unsafe [5]. Most of the harvesting techniques are planned to be self-supporting, cost-efficient, and need no service for several years.

Figure 10.15 shows general Energy Harvesting system which incorporates Transducer, Rectifier, and Regulator [6, 7]. Transducer is an electronic instrument that accumulates energy from any source and changes to electrical energy. Typical transducers are photovoltaic, thermoelectric, piezoelectric, electromagnetic, and RF [8–12, 30]. The output of the transducer is in the form of ac but in order to power electronic devices dc source is required hence Rectifier is used after Transducer. As rectifier provides pulsating voltage and hence a regulator is used after the rectifier in order to get constant voltage so as to power electronic devices reliably. The total PCE of the energy harvesting system depends on the efficiency of the transducer, rectifier, and regulator [6, 7].

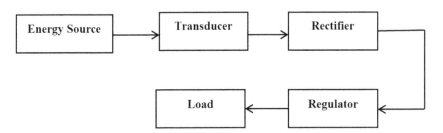

FIGURE 10.15   Block diagram of energy harvesting for IoT.

10.8.1   SIGNIFICANT BASES OF HARVESTING ENERGY IN IoT DEVICES

The basic sources of Energy scavenging techniques are thermal, mechanical, light, RF energy and play a major role to harvest the energy from the environment which are analyzed in the following sections.

10.8.1.1   MECHANICAL ENERGY HARVESTING

Mechanical energy is the summation of kinetic and potential energy in an object due to its movement or position, or maybe of both. Energy scavenging transforms mechanical energy similar to rotation, motions, or vibrations to electrical signals. Among several methods to scavenge mechanical energy, electromechanical transducers are the best developed method. Mechanical energy basic consists of piezoelectric, electrostatics, and electromagnetics transductions [13]:

1. **Piezoelectric:** This effect converts vibration energy to electrical energy. Piezoelectricity refers to internal electrical energy from mechanical power applied, as shown in Figure 10.16. If force is exerted on a piezoelectric substance, the stationary arrangement is distorted, charged particles are moved, and electric power is produced [8–10]. Sources which are commonly used in piezo-electric energy harvesting are vibrations, audio noise, human motion, etc. [14, 15]. Other Energy harvesting applications that do not depend on abrupt mechanical changes utilize piezoelectric membranes and vibrational beams to collect moving oscillations.

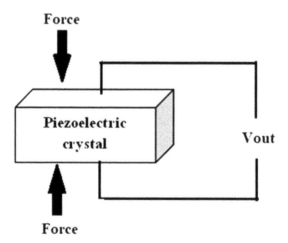

FIGURE 10.16   Piezoelectric effects.

2. **Electrostatic:** This energy process is defined as a capacitive assembly formed by dual individual electrodes as shown in

Figure 10.17. The space in-between electrodes are usually filled by dielectric materials, air medium or vacuum [16–18]. By the movement of any of the electrodes, changes capacitance and this mechanical variation translates to electric energy. Thus, the energy harvested by means of electric field strength is used to activate low energy remote sensor nodes.

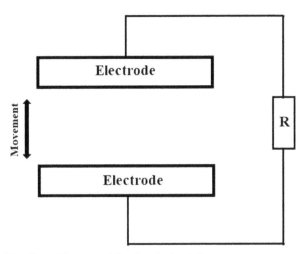

FIGURE 10.17   Capacitive assemblies for electrostatic energy.

3.   **Electromagnetic:** Faraday's law of induction forms the basic method for Electromagnetic energy harvesting technique. An electromagnetic generator has a coil and permanent magnets that induces strong magnetic fields which permits the flow of electric current in one or more direction. One among the coil or permanent magnet should be fixed and the other could be connected to movable frame. Vibration creates relative movement and hence transduction starts functioning and electric energy is produced [19–21]. Figure 10.18 shows generally appreciated models of electromagnetic generators.

10.8.1.2   LIGHT ENERGY HARVESTING

Sunlight is richest renewable energy source. Solar power consumed by earth in a year is far less than that of the energy it receives within an hour. The proceeding topics explain how light energy could be changed to electrical energy:

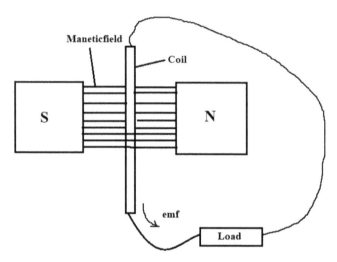

FIGURE 10.18   Electromagnetic energy generator.

1. **Solar Cell:** PV cells are called solar cells, is an electrical device which changes light energy to electric energy using photovoltaic principle. This effect denotes particles of light, i.e., photons stimulating electrons to a high level of energy, permitting them to act as charge particles and produce electrical current [22]. The photovoltaic cell comprises dual layers called N-type and P-type semiconductor and once these two semiconductors are located inside a photovoltaic cell, the free electron from N-type silicon move towards P-type material to fill up gaps in P-type semicon-ductor. So, N-type material turns into positive charge and P-type material is negatively charged. Thus, an electrical field is produced and electricity is delivered shown in Figure 10.19.

2. **LED and Photodiode:** Low-cost photodiodes and Light-emitting diodes are used to scavenge light energy and yield electricity instead of using expensive PV cells. A LED is a light source that radiates light while current runs over it. Electrons in the semiconductor combine with holes, producing energy in the form of photons. A photodiode converts light back to electrical current. When photons are absorbed, current is generated in the photodiode. Less voltage

moves between leads. This electricity could be scavenged and utilized to generate little microwatt power, which is enough for low power applications, like IoT devices.

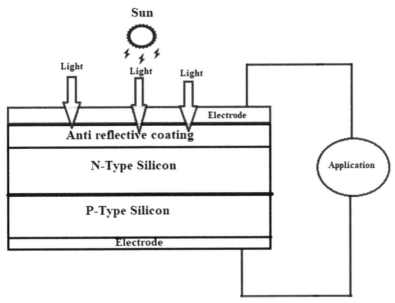

FIGURE 10.19 Solar cell assemblies.

10.8.1.3 *THERMAL ENERGY HARVESTING*

Thermoelectric effect and pyro electrics are the basic techniques used in thermal energy scavenging. Practically all type of transport, production or manufacturing industries, humans, etc., loses heat as waste energy. The harvested energy density differs based on applications and the estimated range is between 2–600 $\mu W/ cm^2$:

1. **Thermoelectric Effect:** The interaction between heat and electricity is known as thermoelectric effects. Thermo energy scavenging is the technique to transform temperature gradients to electric energy by thermopower generators. This thermoelectrical generator is made up of several thermocouples, of N-type and P-type semiconductors which are bounded electrical serially and thermal parallelly. While exposing to thermal gradient, charged particles with negative charge or positive charge will move from

top hot-side to bottom cold-side and electric current is induced as a result of moving charged particles shown in Figure 10.20.

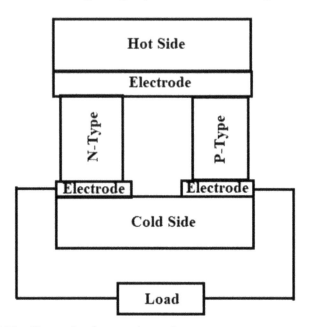

FIGURE 10.20   Thermoelectric energy harvesting.

2. **Pyroelectric Effect:** Basically, a thermoelectric material produces electrical energy from thermal gradients, whereas pyroelectric constituents generate energy from temperature variations, that produce a charge variation on the pyroelectric crystal surface, which generates an equivalent electrical power. Also, this process has certain similarities like piezoelectric harvester that changes mechanical fluctuations to electricity, but this technique can be used only for limited applications because it still needs a few more improvements.

10.8.1.4   *RADIO FREQUENCY ENERGY HARVESTING*

The Radio Frequency is from three kilohertz to 300 gigahertz. Every urban area is enclosed by mobile telephone services, TV broadcasts, radio, and wireless LAN that makes RF emission very effective power source. Due to the RF energy availability and ease of its implementation, wireless energy harvesting is one of the most proven solutions. It receives the transmitted RF waves and converts it into a stable energy source that supplies to the sensor device [12, 15, 24] is shown in Figure 10.21.

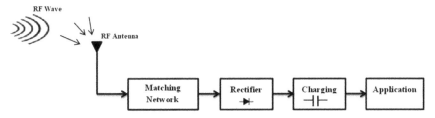

FIGURE 10.21   RF energy harvesting.

10.8.1.5 WIND ENERGY HARVESTING

By using mechanical turbines, wind energy is produced from wind. The turbines use two or three propeller-like blades to capture the wind energy. The turbines are usually kept on tall towers to seize less turbulent, strong wind. The speed of the wind decides the economy and efficiency of wind energy application. High wind revolves blade and spins the turbine shaft that yields electricity. Turbine power generated is proportional to the cube of high wind speed.

10.8.2 APPLICATIONS OF ENERGY HARVESTING TECHNIQUE

Heat, RF energy, Wind energy, Sound, Light, etc., are some different forms of energies that exist in the environment around us. And most of this surrounding energy gets wasted. This type of unusable and unwanted energy will be changed to usable electrical energy to power up electronic devices. Table 10.2 shows power generation of various energy harvesting sources, among them piezoelectric energy harvesting method yields high power density.

10.8.2.1 ENERGY HARVESTING TECHNIQUE (EHT) IN TIRE PRESSURE MONITORING SYSTEM

Due to the rapid growth of the automobile industry, the vehicle running on the roads are increasing day by day. And the numbers of accidents also increase multifold. One among the reasons for road accidents is due to the inflation or burst of tires. Tire burst is a primary botheration for drivers, as it is difficult to predict. The main cause is due to irregular tire pressure and high tire temperature. Inflation pressure is one huge factor that determines performance, impacts tire's speed, and load capacity, cornering power, life service, handling response and overall its protection. Low inflation causes an increase in fuel consumption and sustained will cause structural impairment. TPMS installed in vehicles will measure the inflated tire pressure or change in the tire pressure for a period of time and transmits respective information to the driver when the vehicle is moving. Thus, accidents can be avoided if tire pressure is monitored frequently while driving. Also, TPMS will require changing or maintaining sensors owing to damage or battery fault. Hence to provide less monitoring and battery free sensor application growing interest is shown to use power harvesting techniques to energize TPMS as presented in Figure 10.22.

TABLE 10.2 Power Generation from Various Energy Harvesting Sources

Method of Energy Harvesting	Power Density
Solar cells (outdoor)	15.48 mW/cm^2
Solar cells (indoor)	10.27 µW/cm^2
Piezoelectric	330.19 µW/cm^3
Vibration	115.86 µW/cm^3
Thermoelectric	39.96 µW/cm^2
Wind (outdoor)	3.49 mW/cm^2
Radio frequency	0.9 µW/cm^2

Few piezoelectric sensors with transducer modules are kept all along the inner circumference of the tire. When the tire rotates, it experiences a mechanical stress which is also captured by the transducer module. The transducer module converts mechanical stress to electrical power and fed to capacitor bank for storage.

10.9   THE RANKING OF IoT APPLICATIONS

Several IoT applications exist nowadays with diverse roles to play in our day-to-day life [23, 25]. The ranking of a few of the applications are shown in Figure 10.23.

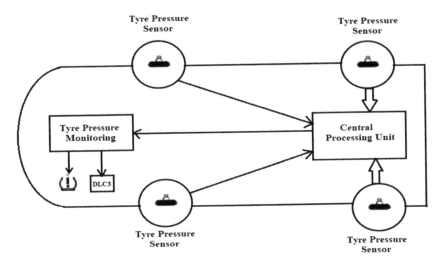

FIGURE 10.22   EHT used in tire pressure monitoring system.

10.9.1   HOME AUTOMATION

This stands out to be the highest IoT application. Every month almost 60,000 people Google the term "Smart Home." More corporations are actively involved in developing smarter homes than other IoT applications. Total funding for Smart house startups exceeds $2.5 billion.

10.9.2   WEARABLE

Wearable is still a hot area to work on IoT. There are plenty of innovations, like smartwatches, trainers, bracelets, gesture control [29], etc. Among all IoT startups, Jawbone is the biggest funding company till date and has invested near to half billion US dollars.

10.9.3   SMART CITIES

There are variety usages, of smart city from road traffic controlling to environmental monitoring, garbage management, water distribution, urban safety, etc. Smart city solutions are very popular, as the solution alleviates most of the real time problems of people in the city. IoT solutions also help in reduction in noise, pollution, and keep cities safer.

10.9.4   SMART GRIDS

Smart grids are distinct ones. Smart grid assures providing useful information about the behavior of electric consumers and suppliers in an automated method with an improvement in reliability, efficiency, and financially.

10.9.5   INDUSTRIAL INTERNET

Even though the popularity of the industrial internet has not reached the mass like wearables or smart home has, this IoT concept has a very high overall potential. Compared to other non-consumer IoT, industrial internet gets the huge popularity through social media.

10.9.6   CONNECTED CAR

The typical development cycle in the automotive industry is 2–4 years and due to this fact, the connected car concept takes some time to come up. Most automakers and some startups are working on connected car solutions. Hi-tech giants like Microsoft, Google, and Apple already announced working on connected car platforms.

10.9.7   CONNECTED HEALTH

Connected health is the sleeping giant of IoT applications. A connected health care system with smart medical devices can provide huge potential, not only for companies but also in general for people well-being. Large-scale successes with prominent cases are yet to be seen.

10.9.8   SMART RETAIL

The popularity ranking shows smart retail still in a niche segment. And the subset of smart retail, proximity-based advertising has started to take off.

10.9.9 INTELLIGENT SUPPLY CHAIN

Logistics chains are already smarter. Solutions like tracking products when moving, supplier inventory information, etc., are already in the market. So, it is perfect to believe that smart supply chain will get a new identity with the Internet of Things, even though its popularity so far remains limited.

10.9.10 SMART FARMING

Smart farming is often looked over business-case for IoT as it does not fit into typical categories like industrial, mobility or health. IoT can easily revolutionize the work of farmers, specifically in areas like remote farming, number of livestock, farming operations, etc. [27]. Smart farming will be a very important application in agricultural-product exporting countries (Figure 10.23).

10.10 FUTURE ENHANCEMENTS OF IoT

10.10.1 EFFICIENT PRODUCTION BY DEVICE MONITORING AND PRODUCT-QUALITY

Machines are supervised and analyzed continuously in order to ensure their performance within the tolerance level. Identification and detection of product quality defects can be monitored real-time.

10.10.2 IMPROVEMENT OF TRACKING AND "RING-FENCING" IN PHYSICAL ASSETS

Asset location could be easily determined by tracking. Ring-fencing ensures protecting high-value assets from being removed or stolen.

10.10.3 MONITORING ENVIRONMENTAL CONDITIONS AND HUMAN HEALTH ANALYTICS USING WEARABLES

IoT wearable ensure people to get a better understanding of their individual healthiness and also permit doctors to observe patients remotely.

IoT enhances corporations to track well-being and care of respective staffs working in hazardous conditions.

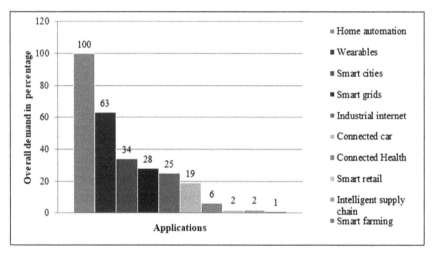

FIGURE 10.23   IoT application analysis.
Source: Padraig Scully, Ref. [32].

10.10.4   *ENABLE PROCESS CHANGES IN BUSINESS*

Monitoring the condition of remote machines and activating preventive measures. Remote monitoring also enables product-as-a-service business methods in which customers pay only for its usage instead of buying a product. IoT usage increases effectiveness and protection in fleet control management.

10.11   CONCLUSION

The summary of this work is to find out the potential of Energy Harvesting techniques in order to minimize or even eliminate the use of batteries in IoT devices. Hence it is appropriate to find the right Energy Harvesting techniques for IoT systems. The most commonly used ambient energy sources, which include mechanical vibrations, wind, RF, solar, and thermal energy. At present, industry is developing energy harvesting technology to extract the range of energy from 10 μW to One W, since IoT systems are mostly in this mentioned power range. Above mentioned energy sources have their own merits and demerits and hence remain a challenge for harvesting energy continuously. Due to this challenge, irregular power urge of loads will occur, which requires sourcing of extra energy in order to fulfill inadequate harvested power. As a future enhancement, this energy insufficiency can be overcome by hybrid energy harvesting technique.

KEYWORDS

- **energy harvesting technique**
- **IoT application**
- **IoT growth**
- **power generation**
- **sensors**

REFERENCES

1. Kaivan, K., (2013). The role of sensor fusion and remote emotive computing in the Internet of Things. *Freescale Semiconductor, 1*, 1–14.
2. Tingwen, R., Zheng, J. C., & Meiling, Z., (2017). Energy-aware approaches for energy harvesting powered wireless sensor nodes. *IEEE Sensors Journal, 17*, 2165–2173.
3. Lee, S. Y., Hong, J. H., Hsieh, C. H., Liang, M. C., & Kung, J. Y., (2013). A low power 13.56 MHz RF frontend circuit for implantable biomedical devices. *IEEE Trans. Biomed. Circuits Syst., 7*, 256–265.
4. Indrajit, S., Sagar, M., & Kalyan, B., (2017). *A Review of Energy Harvesting Technology and its Potential Applications, 4*, 33–38. International Information and Engineering Technology Association.
5. Sankman, J., & Ma, D., (2015). A 12 µW to 1.1 mW aim piezoelectric energy harvester for time-varying vibrations with 450 nA IQ. IEEE *Transactions on Power Electronics, 30*, 5665–5680.
6. Shuenn-Yuh, L., Zhan-Xian, L., & Chih-Hung, L., (2019). Energy harvesting circuits with a high efficiency rectifier and a low temperature coefficient bandgap voltage reference. *IEEE Transactions on Very Large Scale Integration Systems, 27*, 1760–1767.
7. Hsieh, C. H., Du, C. Y., & Lee, S. Y., (2014). Power management with energy harvesting from a headphone jack. *Proc. IEEE Int. Symp. Circuits Syst., 1*, 1989–1992.
8. Ramadass, Y. K., & Chandrakasan, A. P., (2010). An efficient piezoelectric energy harvesting interface circuit using a bias-flip rectifier and shared inductor. *IEEE J. Solid-State Circuits, 45*, 189–204.

9. Lu, S., & Boussaid, F., (2015). A highly efficient P-SSHI for piezoelectric energy harvesting. *IEEE Transaction on Power Electronics, 30,* 5364–5369.

10. Krihely, N., & Ben-Yaakov, S., (2011). Self-contained resonant rectifier for piezoelectric sources under variable mechanical excitation. *IEEE Transaction on Power Electronics, 26,* 612–621.

11. Xiaofeng, L., & Vladimir, S., (2014). Modeling piezoelectric energy harvesting in an educational building. *Energy Conversion and Management, 85,* 435–442.

12. Manal, M. M., Ghazal, A. F., & Abdel-Rahman, A. B., (2018). High-efficiency CMOS RF-to-DC rectifier based on dynamic threshold reduction technique for wireless charging applications. *IEEE Access, 6,* 46826–46832.

13. Harb, A., (2011). Energy harvesting: State-of-the-art. *Renewable Energy, 36,* 2641–2654.

14. Bala, I., Bhamrah, M. S., & Singh, G., (2017). Capacity in fading environment based on soft sensing information under spectrum sharing constraints. *Wireless Networks, 23,* 519–531. Springer.

15. Beeby, S. P., Tudor, M. J., & White, N. M., (2006). Energy harvesting vibration sources for microsystems applications. *Measurement Science and Technology, 17,* 175–195.

16. Bala, I., Bhamrah, M. S., & Singh, G., (2017). Rate and power optimization under received-power constraints for opportunistic spectrum-sharing communication. *Wireless Personal Communications, 96,* 5667–5685. Springer.

17. Miao, P., Mitcheson, P. D., Holmes, A. S., Yeatman, E. M., Green, T. C., & Stark, B. H., (2006). MEMS inertial power generators for biomedical applications. *Microsystem Technologies, 12,* 1079–1083.

18. Kiziroglou, M. E., He, C., & Yeatman, E. M., (2009). Rolling rod electrostatic micro-generator. *IEEE Transactions on Industrial Electronics, 56,* 1101–1108.

19. Bala, I., Bhamrah, M. S., & Singh, G., (2019). Investigation on outage capacity of spectrum sharing system using CSI and SSI under received power constraints. *Wireless Networks, 25,* 1047–1056. Springer.

20. Cepnik, C., Lausecker, R., & Wallrabe, U., (2013). Review on electrodynamics energy harvesters-a classification approach. *Micro Machines, 4,* 168–196.

21. Saswat, K. R., Banee, B. D., Ayas, K. S., & Kamala, K. M., (2019). Ultra-low-power solar energy harvester for IoT edge node devices. *IEEE International Symposium on Smart Electronic Systems.* doi: 10.1109/iSES47678.2019.00053.

22. Sherali, Z., Faisal, K. S., Anum, T., & Quan, Z. S., (2020). Design architectures for energy harvesting in the Internet of Things. *Renewable and Sustainable Energy Reviews, 128,* 1–8.

23. Mahmuda, K. M., Rokonuzzaman, M., Jagadeesh, P., Mohammad, S., Kazi, S. R., Fazrena, A. H., Sieh, K. T., & Nowshad, A., (2020). Prospective efficient ambient energy harvesting sources for IoT-equipped sensor Applications. *Electronics, 9,* 1–22.

24. Alphonsos, A. M., Yan, C. W., & Kok, T. L., (2018). Miniature high gain slot-fed rectangular dielectric resonator antenna for IoT RF energy harvesting. *AEU-International Journal of Electronics and Communications, 85,* 39–46.

25. Gaofei, S., Xiaoshuang, X., & Xiangping, Q., (2019). Energy harvesting-based data uploading for Internet of Things. *EURASIP Journal on Wireless Communications and Networking., 153,* 1–13.

26. Amar, K., Andromachi, T., Piero, G. V., & Sandro, C., (2020). Humidity sensors for high energy physics applications: A review. *IEEE Sensors Journal, 20,* 10335–10344.

27. Himanshu, A., Ruchi, D., Iyer, K. S. S., & Vijayalakshmi, C., (2020). An improved energy efficient system for IoT enabled precision agriculture. *Journal of Ambient Intelligence and Humanized Computing, 11,* 2337–2348.

28. Omar, A. S., Amer, A., Imran, K., & Bong, J. C., (2020). A hybrid energy harvesting design for on-body Internet of Things (IoT) networks. *Sensors, 20,* 1–17.

29. Himanshu, S., Ahteshamul, H., & Zainul, A. J., (2019). Research issues in energy harvesting Internet of Things. *International Conference on Power Electronics Control and Automation.* doi: 10.1109/ICPECA47973.2019.8975520.

30. Abdul-Qawy, A. S. H., Nasr, M. S. A., Srinivasulu, T. A., (2020). Classification of energy-saving techniques for IoT-based heterogeneous wireless nodes. *Procedia Computer Science, 171,* 2590–2599.

31. Jaswal, K., Choudhury, T., Chhokar, R. L., & Singh, S. R., (2017). Securing the Internet of Things: A proposed framework. *2017 International Conference on Computing, Communication and Automation (ICCCA),* pp. 1277–1281. doi: 10.1109/CCAA.2017.8230015.

32. Padraig, S., (2020). *Top 10 IoT Applications in 2020.* IoT Analytics, https://iot-analytics.com/top-10-iot-applications-in-2020 (accessed on 16 November 2021).

33. Sani, T., (2018). *Importance of the Sensors in the Internet of Things.* Electronicsforu. https://www.electronicsforu.com/technology-trends/tech-focus/iot-sensors (accessed on 16 November 2021).

CHAPTER 11

IoT-Based Peltier Air Conditioner

VANKADARA SAMPATH KUMAR,[1] B. PRAVEEN KUMAR,[1] and
CH. SANTHAN KUMAR[2]

[1]*Bharat Institute of Engineering and Technology, Hyderabad,
Telangana, India, E-mail: sampath.vankadara62@gmail.com
(V. S. Kumar)*

[2]*Lords Institute of Engineering and Technology, Himayath Sagar,
Hyderabad, Telangana, India*

11.1 INTRODUCTION

To safeguard the environment, a modification is required to the present air conditioners. A typical ordinary cooling framework [1] has three basic parts: those are, condenser, blower, and evaporator. The pressurized cooling is permitted to extend, bubble, and dissipate in the evaporator part. During this modification of state from fluid to gas, vitality (heat) will be generated.

To recompress the gas to liquid the compressor acts with the refrigerant pump. The absorbed warmth within the evaporator plus the produced warmth during compression, ejected into the environment or ambient by the condenser. At the new junction, energy is expelled to a sink as electrons move to a lower energy element (p-type) from a high energy element (n-type). An analogous thermoelectric part which has the cold side, energy (heat) is absorbed by electrons as they pass from energy within the semiconductor element, to a better energy within the semiconductor unit element. The supply will take care of the energy to control and transfer

Harnessing the Internet of Things (IoT) for Hyper-Connected Smart World.
Indu Bala and Kiran Ahuja (Eds.)
© 2023 Apple Academic Press, Inc. Co-published with CRC Press (Taylor & Francis)

the electrons through the system. For another application like solar still, Peltier can be used [2].

An air conditioner that is very reliable and efficient in the present days has observed some demerits. For the last two decades, the ozone layer is getting destroyed with the release of fluorinated hydrocarbons used for the air conditioning and refrigeration purposes also leaked and slowly ascend into the atmosphere. It has created a threat not just to maintain the earth's ecosystem stable but also because of the existence of the earth. An aim to reduce that, this chapter came up with a new cooling framework utilizing the Peltier Module which will overcome all the hindrances of the existing HVAC framework. Ordinary blower run cooling gadgets have a large number of disadvantages with the utilization of CFC refrigerants with an impact on global temperature alteration. A large portion of the powerage depends on the coal power plants, which add ozone depleting substances to the environment is the significant reason for an unnatural weather change. Peltier coolers have numerous advantages and applications over conventional cooling gadgets, lesser size, light in weight, high undependability, no moving parts or mechanical parts, working liquid and it is working by the principle of thermoelectric effect. Peltier modules have the ability to give the maximum cooling, here in this work the obtained maximum performance is 17.1°C further which is controlled by an IoT device for remote operation. The block diagram of the proposed work is shown in Figure 11.1.

11.2 THERMOELECTRIC EFFECT

With the help of thermocouple, direct conversion of temperature differences to electric voltage and vice versa is known as the thermoelectric effect. When there is a different temperature on both sides the thermoelectric device creates a voltage. Conversely, heat is transferred from one side to the opposite, when a voltage is applied to that which creates a temperature difference. An applied voltage gradient causes a charge carrier within the material to diffuse to the cold side from the new side at the atomic scale.

11.2.1 PELTIER ELEMENT

P-type and N-type bismuth telluride materials were used for the fabrication of semiconductor materials of the Peltier element used in the internal structure. To maximize the thermal transfer between cold and hot ceramic surfaces of the module pellets which are in array electrically connected in series, but thermally arranged in parallel [3]. The internal structure of the generic Peltier element is shown in Figure 11.2.

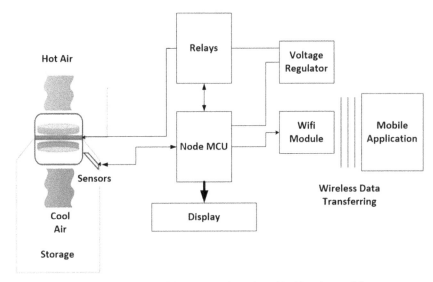

FIGURE 11.1   Block diagram of the proposed IoT-based Peltier air conditioner system.

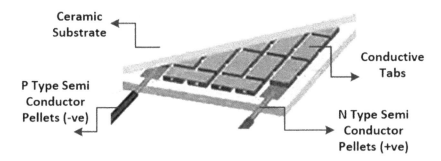

FIGURE 11.2   Internal structure of generic Peltier element.

When a current is passed between two dissimilar junctions heat is being either emitted or absorbed which is known as a thermoelectric effect. For its cooling property, it is used in the Peltier element, which comprises a thermoelectric module with two ceramic plates sandwiched for its high thermal conductivity [4], with an influence source, is effectively able to pump heat across the device to opposite plate from one ceramic plate. Besides, by reversing the direction of current flow, the direction of warmth flow is changed.

The positive and electric charge carriers to immerse up heat from one substrate surface and transfer and release it to the substrate on the alternative side caused by applying a DC voltage. Therefore, the opposite surface, where the energy is released, becomes hot and the surface where energy is absorbed becomes cold.

11.3   WORKING MODEL

Peltier effect operates in line with the thermoelectric coolers. By heat transferring between two electrical junction's temperature difference has been created. At the junction of the conductor's voltage is applied to generate electrical current. When the current flows through the junctions of two conductors, cooling occurs at one junction and heat is removed and deposited at the opposite junction. Peltier effects most application is cooling; however, it may be used for control of temperature or heating [7].

11.3.1   HARDWARE MODEL

The following are the components used in the project:

- Peltier module;
- Node MCU;
- LM35 temperature sensor;
- Exhaust fans;
- Aluminum heat sink;
- Regulated power supply.

11.3.1.1   ESP8266 NODE MCU WI-FI DEVKIT

A microcontroller named ESP8266 is designed by Espressif Systems. Applications that run independently with Wi-Fi will be having added advantage with the support of ESP8266 [5] and shown in Figure 11.3(a). The specifications of the module are given in Table 11.1. ESP8266 module accompanies a rich collection of pin-outs with an inbuilt USB connector and miniaturized USB link with scale, Node MCU can be interfaced with devkit to your PC and glimmer it with no difficulty, similar to Arduino. With additional breadboard which is well disposed of.

TABLE 11.1 Experimental Range and Levels of Independent Variables

Parameter	Specification
Voltage	3.3 V
Flash memory attachable	16 MB max (512 K normal)
Wi-Fi Direct (P2P)	Soft-AP
Processor: 10 silica	L106 32-bit
Current consumption	10 uA ~ 170 mA
RAM	from 32 K–80 K
Processor speed	80 ~ 160 MHz
Analog to digital	Input 1,024 step resolution
GPIOs	17 pins (multiplexed with other functions)
Maximum concurrent TCP connections	5
802.11 support	b/g/n

11.3.1.2 TEMPERATURE SENSOR (LM35)

Proportional characteristics with Celsius (centigrade) to output voltage present in the LM35 temperature sensor series with at most precision is used here and shown in Figure 11.3(b).

11.3.1.3 ALUMINUM HEAT SINK

An aluminum heat sink is a heat-absorbing metal gadget that will recollect the scatter heat away from a high-temperature heat article, like a PC processor and worked to keep both the CPU and the warmth sink at a suitable temperature. With the help of metal, warmth sinks are made, for example, a copper or aluminum composite, and are joined to the processor [5]. Most warmth sinks have blades, dainty cuts of metal associated with the base of the warmth sink, which assists spread with warming over an enormous zone.

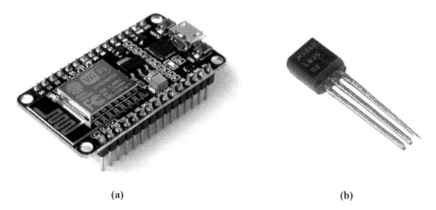

(a)   (b)

FIGURE 11.3   (a) NodeMCU; (b) temperature sensor LM35.

Warmth sinks are generally used in all CPU's and are used in refrigeration and cooling frameworks, GPUs, and video card processors. A conductor could be a latent gadget that moves the glow formed by a robot or an electronic to a liquid moving which is coolant. By at that moment moved warmth with the smooth development of a device, thus allowing the rule of the contraption temperature at really conceivable levels. A conductor is expected to enlarge its area in contact with the cooling medium surrounding it, similar to the air. Speed of air, the decision of texture, bulge structure and surface treatment are industrial facilities that influenced the presentation of a conductor. Copper is used to make the conduit material or potentially aluminum. Aluminum is used in applications where weight could be a major concern.

11.3.1.4   FANS AND EXHAUST FANS

Fans all perform the fundamental function of moving air from one space to a different. But the nice diversity of fan applications creates the requirement for manufactures to develop many various models each model has performing the air movement function. The trick for many users is sorting through all of the models available to seek out one that is suitable for anyone's need; here are some guidelines. The used fan model is shown in Figure 11.4.

It is a follower which is employed to get rid of moisture out of space. It helps to get rid of any odors. The first purpose of the fan is to manage the inside environment by venting out smoke and other contaminants that can be present within the air. It will be integrated into a cooling or heat. It disperses the air harmlessly. It will be utilized in summers to push warm air out for temperature control. It will be used as an alternative for cooling.

FIGURE 11.4 Exhaust fan used.

11.4 SOFTWARE MODEL WITH MOBILE APPLICATION

11.4.1 FLASHING NODE MCU FIRMWARE ON THE ESP8266 USING WINDOWS

ESP8266 modules with content of LUA is programmed by the firmware node MCU fundamentally similar to the way we program to Arduino. The only difference is the couple of lines to give code that builds up to control the ESP8266 GPIOs, to transform ESP8266 to a web server and to build up a Wi-Fi association and many more.

11.4.2   BLYNK APPLICATION

To operate the things using the Internet of Things Blynk application is the best suitable platform which can have gears control by remotely, sensor data visuals, to do various things, forecast the things and many more at low cost.

There are three important parts in the application categorized as:

1. **Blynk App:** Permits interfaces to your gadgets for controlling, utilizing, performing with utmost control.
2. **Blynk Server:** Responsible for all the interfaces between the equipment and cell phone for storage and command prompt we can utilize the Blynk cloud and local space in the Blynk server. With its cloud application it can run a large amount of data with several gadgets also can be applicable to run on a Raspberry Pi for extended application.
3. **Blynk Libraries:** To execute all the procedure correspondence to the commands from the server and sequence operation of conventional equipment with stages.

If anyone visualize the Blynk application performance that every time a person hit a button in the app a command goes to the concerned gadget connected to it through the server of Blynk cloud and the performance is seen in the gadget accordingly in a Blynk of the eye shows the beauty of the application.

11.4.3   FEATURES OF THE APPLICATION

The following are the key features of the application developed. The block diagram of the Blynk application interface is shown in Figure 11.5.

- Connection to the Blynk cloud using:
  - o   Bluetooth and BLE;
  - o   Wi-Fi;
  - o   USB (serial);
  - o   Ethernet;
  - o   GSM.
- Pin manipulation directly with no code writing;
- For all supported hardware and device similar API and UI;
- History data monitoring through super chat widget;

- Set of easy-to-use widgets;
- Device-to-device communication using bridge widget;
- Sending emails, tweets, push notifications, etc.;
- Easy to integrate easily also can add new functionality by using virtual pins.

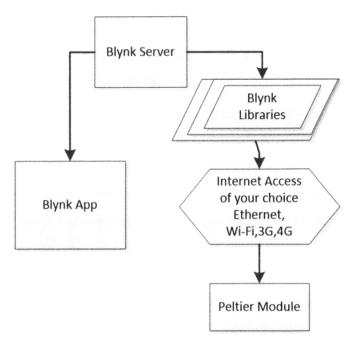

FIGURE 11.5 Blynk application interface.

11.5 RESULTS

The working on the proposed work t is the same as shown in the block diagram and hardware experimental setup is shown in Figure 11.6. As the parameters like current, voltage, and quality of the Peltier module can vary the temperature from time to time. The desired output was 16°C but the recorded output is from 17.1°C to 24°C. Hence, it is concluded that the design of the system and the cabinet to be effective for getting the desired cooling.

Figure 11.6 shows the working and the arrangements of the hardware appropriately.

- Here, the RPS gives the 12 V supply to both the relays;
- One relay is connecting the fans and the other to the Peltier modules;
- A temporary battery is used for the Peltier modules;
- There is Node MCU for wireless communication and wireless control;
- There is a temperature sensor to show the current temperature;
- This wireless data transferring and control in done over WI-FI through a mobile application called Blynk.

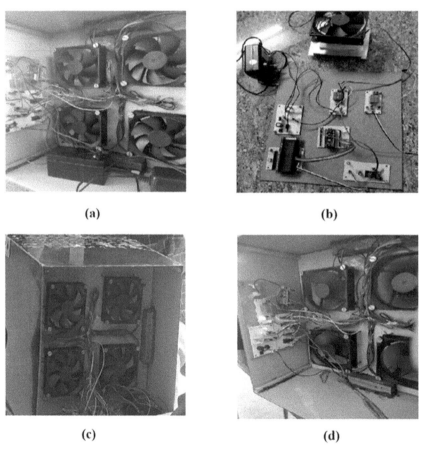

(a)　　　　　　　　　　(b)

(c)　　　　　　　　　　(d)

FIGURE 11.6 Hardware setup of the proposed IoT-based Peltier air conditioner (a) internal setup; (b) view of IoT controller circuit; (c) full view of the setup; (d) working of the proposed system.

11.5.1 APPLICATION RESULTS

Figure 11.7 is a screenshot of mobile application results with Blynk mobile application and the results are shown:

- Here, two parameters, i.e., input voltage and temperature were shown;
- From there, the system can be operated and controlled;
- An On/Off button is given to switch on and off the system;
- And a speed slider is given to control the speed of the fans;
- There is a mobile alert icon that shows or alerts when any defects or failures occur.

The maximum cooling obtained during the peak performance is 17.1°C and can be varied up to 24°C at 12 V Supply, 0.5 Ampere, when the room temperature is at 30°C, it can be improved that the design of the system and the cabinet need to be effective to obtain the desired cooling. All the operations and controls are done wirelessly over Wi-Fi through the Blynk application and the output in mobile application is shown in Figure 11.7.

11.6 CONCLUSION AND FUTURE SCOPE

11.6.1 CONCLUSION

Since Peltier cooling is not productive similarly and because of its little size applications, it is not generally utilized. It discovered its application just in hardware cooling and so on. However, it is seen that there is an enormous extent of research right now thermoelectric materials, its manufacture, heat sink plan and so forth. With the differences between second law skill and first law effectiveness, scientists are taking a shot at diminishing irreversibility in the frameworks, that Peltier cooler has tremendous potential with improved cooling.

This work aims to achieve cooling in the long term for remote areas to use portably, affordable to needy people in case of power cuts with an alternative to the refrigerator which can provide good cooling with an input of 12 V module. The first cooling calculation was considered under-study with the cooling of 24°C. With the full performance and power, it was managed to bring the cooling to 17.1°C. The time taken to reach from room temperature to the maximum cooling was 13 minutes.

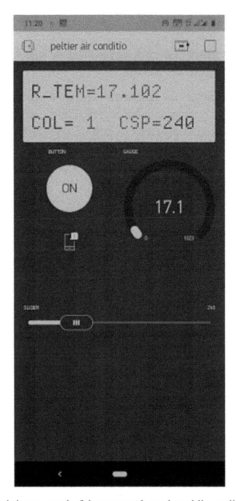

FIGURE 11.7　Real time control of the system through mobile application.

11.6.2　FUTURE SCOPE

With the recent production of different materials having thermoelectric properties with high-temperature variation to be explored in the field of thermoelectric and nanoscience, which helps further temperature cooling, lower current consumption at the higher ambient day to day conditions. Reliability of power in thermoelectric applications, fast mechanisms need to be explored.

The present project can be developed to a sustainable system where solar panels are used for power supply, and it can be used by the people who are staying at a remote location where there is no electricity. It can also use this in medical and pharmacy applications where the organs, blood, and medicine which need a particular temperature to protect it and the whole storage system can be attached to the drone and transfer to the required location.

KEYWORDS

- **Internet of Things (IoT)**
- **Peltier air conditioner**
- **Peltier cooler**
- **Peltier module**
- **thermoelectric**

REFERENCES

1. Julian, G. H., (2014). Bismuth telluride and materials for thermoelectric generation. *Materials, 7,* 2577–2592.
2. Pounraj, P., Prince, W. D., Kabeel, A. E., Praveen, K. B., Muthu, M. A., Ravishankar, S., & Cynthia, C. S., (2018). Experimental investigation on Peltier based hybrid PV/T active solar still for enhancing the overall performance. *Energy Conversion and Management, 168,* 371–381.
3. Stackhouse, S., & Stixrude, L., (2010). Theoretical methods for calculating the lattice. *Thermal Conductivity of Minerals, Mineralogy & Geochemistry, 71,* 253–269.
4. Melcor homepage http://www.melcor.com (accessed on 16 November 2021).
5. Riffat, S., Omer, S. A., & Xiaoli, M., (2001). Thermoelectric refrigeration system employing heat pipes and a phase change. *Renewable Energy, 23,* 313–323.
6. Chein, R., & Chen, Y., (2005). Performances of thermoelectric cooler integrated with microchannel heat sinks. *International Journal of Refrigeration, 28,* 828–839.
7. Riffat, S., & Ma, X., (2004). Improving the performance of thermoelectric cooling systems. *Int. J. Energy Res., 28,* 753–768.

CHAPTER 12

Aspects and Use of Digital Agriculture Using IoT

AYUSH KUMAR AGRAWAL and MANISHA BHARTI

National Institute of Technology Delhi, Delhi–110040, India,
E-mail: ayush6295@gmail.com (A. K. Agrawal)

12.1 INTRODUCTION

Nowadays, the universe is inhabited among about 8 billion people. The figure is projected to hit more than 10 billion in another three decades. Integrate the population explosion with the imminent labor deficit from a solely numerical viewpoint—provided that the world's total populace is aging and fewer people have come into the workforce—the farming industry needs to generate further. It has contributed to the rise of technology and computers in the farming sector, like smart agriculture, which is now essential for the sector's effort to scale up with this increasing market forces. Perhaps one of the most neglected facets of agriculture, when smart agriculture has gone into force, is draining and cultivation because both are highly necessary for seed germination. The soil requires the correct level of fluid to guarantee it does not harm the quality of the seed. This causes erosion as the water rates get too large. This contributes to seed overwatering, causes contaminants to migrate through the environment, etc. By refurbishing storm drains to IoUT sensors, growers are better equipped to closer track groundwater regulation and launch. In fact, such IoUT apps enable growers to automatically track irrigation systems, from comfort and privacy.

Harnessing the Internet of Things (IoT) for Hyper-Connected Smart World.
Indu Bala and Kiran Ahuja (Eds.)

The world population has exploded in volume ever since the Great Agricultural Boom of the 19th century. This has led to increasing competition and stress on the agricultural sector to meet the required product that the new population demands. The integration of IoUT technological advances has been crucial in digitizing the sector and optimizing demand for our up with the fast-growing population. Although the use of IoUT technology inside the agricultural sector is still in the greenbelt, this is already having a significant effect by automating routine farming procedures and high inflation. For example, the two most prominent forms IoUT shapes the sector at the moment are precise agriculture and tracking of livestock. Although IoUT technology would be used over the last several generations of livestock tracking, its usage has been restricted to livestock—poultry, cows, and pigs used to grow food or feed in tight quarters because they are all accessed easily. A radically new area for IoUT technologies has been posed by the latest drive to gather and evaluate information on "farm fresh" poultry, goats, and cows. It poses a different experience in the tracking of livestock—concerning what we are looking to gather data on this sort of animal. For example, questions such as which look like good free-range cattle, how frequently they do feed, what squared region of ground they need to become stable, etc. Furthermore, customers are becoming wiser and mindful of whatever they eat. The typical customer decades ago may not even have understood what free-range poultry or meat was. Modern customers become increasingly personalized with their preferences, and now they want to ensure they become buying the best healthy, healthier items, it was not just one winner take all mindset. This means a number of IoUT devices are required to track livestock's pulse rate, momentum, and position throughout the region. These recent technological advances and increased data pool give people peace of mind—this goat has expected to spend 93% of his period on an old farm that they are seeking to make wise choices.

Besides becoming more competitive and comfortable, IoUT may also increase labor as well as other costs by growing available facilities rates. If the agricultural sector begins to evolve technologically, they can start to see a larger need for simplified IoUT solutions. The criteria for each of these technical approaches are actually diverse-varying from mobile accessibility and health to energy usage. When agricultural operations are increasingly streamlined, the concept of 'co-creation' will start to pick up momentum. This, in effect, would help push the sector's IoUT technologies further, putting combined technological experts from various sectors to develop agricultural-specific solutions that can be customized to customer requirements [1–5].

Linked technologies have invaded every part of our lives with the increasing embrace of the IoUT, from safety and wellness, home security, transportation, to smart buildings, and advanced IoUT. Therefore, it is really only natural that IoUT, technology factor, and technology find their application in the field, and therefore, greatly enhance almost every aspect of it. Over the last generations, agriculture has undergone many technical changes, and become more organized and technology-driven. Farm-workers also acquired greater influence over the cycle of land for growing crops with the use of numerous precision farming devices, rendering it much more efficient and increasing its effectiveness.

12.2 APPLICATIONS OF IoUT FOR DIGITALIZATION IN AGRICULTURE

Easily the most widespread modern agriculture devices are weather fore-casts, which integrate multiple detectors for smart agriculture. They are situated around the ground, collecting different environmental data, and sending it to the data center. The measures given could be used to chart the climate patterns, pick the correct varieties, and undertake the requisite steps to boost their efficiency (i.e., smart agriculture). Weather forecasts may also periodically change the circumstances to suit the specified criteria, in contrast to collecting climate information. In fact, plantation automated machines follow a common concept.

Another form of IoUT product in cultivation is conservation agricul-ture equipment, and then another aspect of smart agriculture. Much like satellite data, they must be installed in the fields to gather crop-specific information; from precipitation and temperature to the capacity for water potential or crop yields safety. Then you can evaluate their agricultural production as well as any abnormalities to possibly eliminate any illnesses or insect infestation that could damage your output. Much like farm tracking, IoUT agricultural devices may be mounted to the cattle on a field to track their safety and output in logging. And this so-called agricultural efficiency-management system will reflect a more complicated approach to IoUT items in farming. Typically, they contain a range of farming IoUT tools and detectors, mounted at the property, and also a strong interface with monitoring capability and built-in financial reporting/development confidence [6–9].

The human market is projected to reach 10 billion inhabitants by 2040 as per the United Nations Food and Agriculture Organization (FAO). In order to provide adequate food to feed in question, the amount of agricultural output would account for more than 40%. Considering that assets for farming production are scarce (almost all of the ground available for farming is already being used), only another way to boost output is by increasing productivity in output. There is little question as to the degree to which precision agriculture will help to overcome this obstacle; however, now without, that does not seem likely.

The invention of reasonably priced IoUT could even modify the irrevocable costs of excessive irrigated areas by giving landowners real-time monitoring of water flow as well as faraway irrigation reporting tools. The sensor-based device has sense levels of water at different depths and optimizes the monitoring of the groundwater valve. LoRaWANTM based learning technologies include manufacturers with actionable information, while allowing for even more water recycling, improved plant production, enhanced agricultural production, and 30 to 40% reductions in agricultural water consumption. Till now, farmers have had no appropriate means of measuring local climate apart from visually inspecting the plants. Now IoUT binds the lands and offers everyone the power to inform you what they want and need to flourish. Devices for soil tracking provide real-time sediment temperature measurements, thermal moisture content, ambient temperature, but also nitrogen, phosphate, and potassium (NPK) thresholds—directly from the sector to the web. Intelligent devices have adjustable cycles of data, removing the need for published studies.

One could forecast optimum plantation and implantation periods correctly, the water intake, and mitigate the tension from over- or underwatering of the crops. IoUT allows simple and cost-effective tools for tracking environmental forecasts, flood prediction, and water quality management, supplying you with the data to make informed decisions regarding your seeds and farms. For your exact position, view real-time water level details in rivers and canals, the runoff, temperature fluctuations, weather patterns, air quality, and humidity. Optimize energy, water usage, and crop safety with the precision agriculture smart approach. Environmental metrics can be assessed in real-time by deploying a network of LoRaWAN sensors and gateways around the fields, monitoring challenges until they become crises. LoRaWAN related IoUT solutions reflect a significant development in ranching. The sensor data allows ranchers the opportunity to track their cattle even more closely and to respond more rapidly in the case of a disorder—either in the safety of a cattle or in its position. Using the Actility IoUT networking framework, ranchers will now use LPWA networks and smart sensors to control animal safety, activity, and position remotely, be alert to early indicators of pet infection and map livestock movements to help manage pasture use [9–13].

Farm institutions have different assets that require supervision. Broad area and reduced power connectivity supported by LPWA services such as LoRaWAN make for new productive tank and reservoir level control solutions. The smart applications and trackers will secure farm gates, buildings, and equipment. Both beehives provide mobile healthcare spying and the defense against robbery. In fact, IoUT allows increased tracking and quality control of agricultural production by the remote control of the position and circumstances of deliveries and goods. Laborious, ineffective, and incorrect monitoring of the fluid level (water, diesel, oil) in containers, wheat thresholds, pressure, and temperature in bunkers is often done. With LoRaWAN detectors, it is now simple to turn mechanical measurements with an intrusion detection system in order, offering your insight at tank and silo rates and the status of stocks. You can know precisely how often corn, wheat, feed, fuel, and water you have on the side at a certain time, allowing users to manage ahead refill timetables and avoid going out of inventory. Installed within beehives, LoRaWAN battery-based sensors enable beekeepers to remotely monitor the hive condition, such as sensors for temperature and humidity. It helps beekeepers to monitor their bee colonies' life cycle remotely, to improve the safety of the bees, to raise excess mortality levels. GPS trackers may also be included in the solution to know where they are located and prevent hive theft. Partner of Actility offers a user-friendly interface that lets you track and interpret the data that you have gathered.

It is impossible to use conventional gate alarms there because of the intermittent mobile service in rural agricultural areas. Smart LoRaWAN sensors may be placed on windows, remote country houses and cottages, transmitting warnings to open and close windows. GPS devices that are enabled by LoRaWAN could be placed on different agricultural equipment, vehicles, and equipment. This means you will protect your fences, set up alarms for geofencing and activity, and prevent thefts of livestock and machinery with an easy-to-deploy IoUT system and decades of auto-mation detectors. After fertilization occurs, LoRaWAN devices detect the pace of shipment and the atmosphere guaranteeing that food is identifiable and held to maximum safety standards. Quality management control is maintained by tilt, temperature, humidity, air pressure, and so on measuring sensors in any phase in the production process. If temperatures fall below a reasonable range, instant warnings will be set. Quality control informa-tion is submitted to the software repository like consistency metrics and made accessible to both suppliers and customers.

But it has been a good amount of time even though devices were introduced in farming activities. The problem with the existing strategy that uses sensor systems, even so, would be that humans just could not get the real-time information from various sensors. The indicators recorded the information into their connected storage so that we could subsequently use it. Much more sophisticated sensors are being used with the advent of Industrial IoUT in agriculture. Cellular/satellite networks also link the devices to the internet. That lets us know the sensors' real-time details, allowing successful decision-making. IoUT technologies in the farming industry have helped farmers track the water tank rates in real-time, making the irrigation cycle more effective. The advent of IoUT technology in agricultural operations has enabled the usage of sensors like how much time and energy a seed requires to become a fully developed crop in any phase of the farming cycle [14–21].

The IoUT in Agriculture has risen as a second Ecological Disrup-tive Surge. The benefits the farmers are obtaining by adopting IoUT are multiple. It has helped farmers reduce their costs and at the same time raise yields by strengthening farmers' decision-making with accurate data.

Smart farming is a high-tech and productive method of sustain-able farming and food rising. It is an aspect of bringing smart tools and emerging technology into agriculture together. Smart farming is largely dependent on IoUT thereby eliminating the need for farmers and growers to do physical work and thus increasing productivity in every way possible. With the latest agricultural developments based on agriculture, the Internet of Things has provided tremendous advantages, such as effec-tive water usage, resource management, and much more. What made a difference were the enormous benefits, and in recent days this has become a revolutionized agriculture. IoUT-based Smart Farming strengthens the whole Agriculture network by real-time field tracking. The Internet of Things in Agriculture has not only saved the farmers' time with the aid of sensors and interconnectivity, but it has also minimized the wasteful usage of resources such as water and energy. It keeps under control various factors such as humidity, temperature, soil, etc., and gives a crystal-clear observation in real-time.

Climate plays a very important role in agriculture. And possessing insufficient information regarding the environment severely deteriorates the crop production quantity and efficiency. But IoUT applications let you learn the environmental conditions in real-time. Sensors are mounted inside and outside the agricultural areas. They gather environmental data that is used to select the best crops that can grow and survive under specific climatic conditions. The whole IoUT ecosystem consists of sensors that can very reliably sense real-time environmental conditions such as precipitation, wind, temperature, and more. Numerous sensors are accessible to detect all these parameters and customize to match your smart farming needs accordingly. These sensors track crop conditions and ambient conditions. If there are some alarming environmental patterns, then an alert is issued. What is removed is the need for physical activity in challenging climatic environments that eventually improves profitability and allows farmers to reap further benefits from agriculture.

Precision agriculture/precision farming is one of Agriculture's most popular IoUT applications. By introducing smart farming applications such as livestock surveillance, vehicle tracking, field inspection, and inventory monitoring, it allows farming activity to be more accurate and regulated. The goal of precision farming is to analyze the data produced by sensors in order to respond accordingly. With the aid of sensors, precision farming lets farmers produce data and evaluate the knowledge to make wise and fast decisions. There are various precision farming strategies such as irrigation management, livestock management, vehicle monitoring, and several others that play a crucial role in improving productivity and effectiveness. With the aid of precision cultivation, soil conditions and other associated parameters can be evaluated to improve organizational effectiveness. Not just that, you can even monitor the linked devices' real-time operating conditions to measure the water and nutrient rates. To render our greenhouses smart, IoUT has allowed weather stations to change the climate conditions automatically according to a particular set of instructions. The adoption of IoUT in Greenhouses has reduced human interference, rendering the whole operation cost-effective while, at the same time, increasing precision. For example, modern, and inexpensive greenhouses are designed utilizing solar-powered IoUT sensors. These sensors collect and transmit the data in real-time and help to track the condition of the greenhouse quite precisely in real-time. Water consumption and greenhouse status can be tracked via emails or SMS warnings with the help of the sensors. Simple and intelligent irrigation is performed with the assistance of IoUT. Such sensors also provide details on the levels of heat, humidity, temperature, and light.

The standard computer device may not provide adequate capacity for the data from the IoUT sensors. Cloud-based data management and an end-to-end IoUT application play a major role in the smart farming framework. Such technologies are expected to play a crucial role in providing for improved practices. Sensors are the primary source of large scale data gathering in the IoUT world. Using analytical software, the data is processed and converted into useful knowledge. The data analytics aid to track temperature, livestock conditions, and crop conditions. The generated data leverages technical advances and so allows informed choices. By collecting the data from sensors, you will know the crops' real-time status with the aid of the IoUT apps. Use predictive analytics, you will obtain knowledge for making smarter harvest-related decisions. Trend forecasting lets farmers learn about future environmental patterns and crop harvesting. IoUT in the Agriculture Sector has helped farmers preserve seed quality and soil productivity, thus increasing the quantity and price of the goods [22–26].

Technological developments have nearly revolutionized agricultural operations and the arrival of agricultural drones is the disturbance of the cycle. Both ground and aerial drones are used for crop health assessment, crop monitoring, planting, crop spraying, and field analysis. With proper strategy and real-time data-based forecasting, drone technology has provided the agriculture industry with a high rise and makeover. Drones with thermal or multispectral sensors recognize the areas where irrigation adjustments are needed. Once the crops begin to grow, the sensors indicate their health and calculate their index of vegetation. Smart drones gradually decreased the effects on the area. The findings were such that the soil was significantly lowered and even smaller contaminants entered.

Precision farming can be described as something that allows agricultural activities to be more regulated and precise when it comes to animal husbandry and crop development. A core component of this farm management strategy is the usage of Software, including different products such as cameras, control systems, robots, autonomous vehicles, advanced equipment, variable rate technology, etc. The manufacturer's adoption of high-speed internet access, mobile devices, and reliable, low-cost satellites (for imagery and positioning) are a few key technologies that characterize the trend in precision agriculture.

Precision agriculture is one of the most popular uses of IoUT in the agricultural field and this method is utilized by various organizations worldwide. CropMetrics is a precision farm organization focused on state-of-the-art agronomic solutions while specializing in precision irrigation management.

CropMetrics' devices and services include VRI optimization, soil humidity probes, computer PRO optimizer, and so on. Optimization of the VRI (Variable Rate Irrigation) maximizes productivity with topography or soil variation on irrigated croplands, boosts yield, and enhances performance in water usage. The soil moisture probe system offers full local agronomy assistance throughout the season, and guidelines for maximizing the performance of water usage. The virtual optimizer PRO integrates various water management technology into one central, cloud-based, and efficient location built for consultants and growers to take advantage of the advantages of precise irrigation with a streamlined interface.

Incorporating drones nowadays, agriculture is one of the main industries. In agriculture, drones are used to improve different agricultural activities. The ways in which crop health evaluation, irrigation, seed inspection, seed harvesting, planting, and soil and field research are used in agriculture are ground-based and aerial mounted drones.

The main advantages of using drones include crop health imaging, automated GIS visualization, ease of usage, time-saving, and yield-increasing capacity. Through policy and preparation focused on the gathering and analysis of real-time data, the drone technology would offer the agriculture industry a high-tech makeover.

Precision Hawk is an agency that utilizes drones to gather useful data through a collection of sensors that are used to photograph, chart, and survey the agricultural property. These drones do monitor and observations in-flight. The farmers join the area to assess data and pick an altitude or land resolution. The greenhouse IoUT sensors have details on the sun, sound, humidity, and temperature rates. Such sensors will automatically regulate the actuators to open a window, to switch on lamps, to trigger a furnace, to switch on a mister or to turn on a fan, all powered by a Wi-Fi signal.

We may derive insights from the drone data on plant health indices, plant counting, and yield estimation, plant height calculation, canopy cover mapping, field water ponding mapping, scouting studies, inventory calculation, wheat chlorophyll measurement, irrigation mapping, weed pressure mapping, and so on. Throughout the ride, the drone captures multispectral, thermal, and visual images, and then lands at the same location it started off.

Large farm owners may use wireless IoUT applications to gather data regarding their cattle 's position, welfare, and fitness. Such knowledge lets us classify injured livestock and they can be isolated from the group and thereby deter disease transmission. This also reduces labor costs, as ranchers may find their cattle utilizing IoUT-based sensors. JMB North America is a company that provides livestock farmers with cow tracking solutions. One of the approaches lets the cattle owners track female and birth-prone cows. When the water splits, a sensor operated by a battery is removed from the heifer. Which gives details to the rancher or the farm boss. The sensor helps farmers to be more concentrated in the time spent with the heifers that are giving birth.

Specific sensors are used to monitor the environment in a smart greenhouse and calculate the environmental parameters according to the plant requirement. Once linked via IoUT we can build a cloud server for remote access to the network. Which removes the need for manual constant supervision. The cloud service also enables data collection inside the greenhouse and imposes a control motion. This architecture provides farmers with minimum manual interference with cost-effective and efficient solutions.

Illuminum Greenhouses is a drip installation and greenhouse company for Agri-Tech which makes use of new, innovative technology to provide services. It develops new, inexpensive greenhouses utilizing IoUT-sensors with solar power. Using these sensors, the greenhouse condition and water consumption can be tracked through an online platform through SMS warnings to the farmer. In such greenhouses, automated Irrigation is done.

12.3　CHALLENGES OF IoUT IN AGRICULTURE

Farmers need to make irrigation systems more effective and reduce water losses. Large farms have complicated irrigation schemes and irrigation processes, sometimes not effective when manual or time-scheduled moisture regulation and valve openings are in lieu on a need-to-do basis. The soil is not homogeneous, so it retains moisture differently in various places and it has to be closely controlled. Connectivity has always also been a challenge for suburban and remote communities owing to the shortage of coverage and the increasingly high wireless service rates [6–10].

Farmers, when operating in their fields, require real-time visibility in soil conditions. The ability to identify and make required improvements from infected plants was always too late and resulted in crop failure. Manual soil tests do not have the evidence growers need during the growing season and this may lead farmers to either over-fertilize or underfertilize, damage income and/or the ecosystem. Growers often depend on erroneous, inaccurate climatology reviews, and temperature apps or manual processes invest long hours verifying rain gages. Waterways and dams pose a risk of waterlogging, and farmworkers need warning signals if it does occur. Hydroponic gardening and fish populations also require real-time and detailed water management. Bovine injuries, robberies, and casualties are very costly to livestock producers. The manual process of the circumstances and health of livestock is a tough and resource-consuming job. Livestock planning is a challenging, all-round job, but no dairy farmer was ever able to be everything at once-so far. In addition, wild animals in mineral resources are fitted with a tracking system for wellness, scholarly or safety objectives: to limit human intervention, the batteries of the detectors must last for at least approximately 10 months, that is only possible with both the energy consumption and geolocation capabilities of the LPWA channels.

12.4   SOLUTIONS OF THE ABOVE CHALLENGES FACED IN IoUT-BASED AGRICULTURE PROCESS

Optimized regulation of irrigation focused on various parameters including soil moisture, water movement, energy consumption, and environmental factors. LoRaWAN sensors have reliable soil water stress reading-an indicator of how much water is accessible to plants in the field. Irrigation pumps are remotely turned on and off via LoRaWAN. To bypass manual switches a wired system may be attached to every current pump. Actility is the perfect IoUT networking tool for network and sensor control, together with specialized apps of the partners. The activity does provide the IoUT connectivity platform for managing a low-cost, wireless, and remote LoRaWAN-based system, offering farmers real-time insight into their crop's soil condition. In the fields, smart battery-powered sensors are mounted to calculate the NPK rates, pH rates, salinity, moisture, temperature, and aeration of the local soil, as well as the overground temperature and humidity. Actility provides farmers with the perfect IoUT networking tool to operate an LPWA network and IoUT sensors to track a variety of environmental variables such as air quality, flood warning water rates, energy temperature, soil humidity, and more. LoRaWAN helps sensors to track battery life in remote areas within years. The cattle are marked on an ear or collar with LoRaWAN-enabled sensors and trackers that provide position and/or health information. Geofencing is used to monitor when cattle have strayed from a designated region when vital signs and other data such as air temperature and humidity can be recorded to assess cattle condition and stress rates. The activity allows LPWA networks to be managed and the solution implemented, with the application of partners. Trackers are also used for wild animals: The triangulation technique based on LoRaWAN allows authorities to locate them in vast areas without having to resort to GPS [9–14].

12.5 BENEFITS OF USING IoUT

With Smart Irrigation IoUT solutions, growers plan and add just the right amount of water to crops as appropriate, so they can minimize water usage dramatically, improve production and benefit by increasing crop yields. LoRaWAN-based networks are suitable for agriculture due to the low cost of transmitting data at very low voltage, with a battery life of up to 15 years of sensors, enabling farmers to scale up their installations and profit from real operational visibility. The intervention of the operator is limited and therefore the farmer has lower maintenance and operating costs. IoUT approach for soil health monitoring helps farmers to track soil quality from top layers to below roots by providing them exposure in real-time. Detailed data on soil quality help farmers minimize waste and increase crop yields, and evaluate past trends to make smarter decisions on long-term crop management. The data can be used by farmers to identify trouble areas and to evaluate soil within zones. Using IoUT tools, you can efficiently and remotely monitor rainfall in different sections of the field, enabling effective irrigation scheduling and daily updating to enable rapid evaluation of rain events. Alarm levels should be adjusted where flood hazards and water quality adjustments are observed, allowing ample room for corrective steps.

IoUT cattle monitoring solution allows for preventing theft and recovering lost cattle, improving the welfare and productivity of livestock. Effective detection and isolation of injured cattle from herds may be carried out easily, thus tracking of the site increases control and care of the livestock. Applications search data for outliers like a cow displaying decreased agility or increased body temperature that may be symptoms of illness. In fact, the geolocation of LoRa TDoA allows for an estimated position of wild animals thus maximizing the battery life of the device.

12.6 FUTURE OF IoUT IN AGRICULTURE

Landowners too have begun hiring predefined high-tech agricultural methods and technologies to improve the strength of their everyday jobs. For instance, sensors installed in crops enable farmers too because of everyone mapping of the area's geography and properties, and parameters such as alkalinity and soil conditions. As well, they could have direct exposure to weather predictions to predict climate patterns in the coming days. Growers could use one's mobile phones to remote locations, monitor their equipment, crops, and livestock, as well as gather data on trying to feed and increase their livestock. We may also use this method, using numbers, to predict their crops and livestock.

But nevertheless, drones were an excellent component for farmworkers to monitor their estate yet gather fieldwork data. Intelligent farming and precise cultivation are gradually taking off, but they would be only the forerunners of much of the farming techniques. The advancement of block-chain technology is making its way to an IoUT, which can be important in the agricultural sector for its ability to offer useful crop data to enterprises. Landowners that use cameras to detect crop information entered on the blockchain, including various maximum and also the quantity of salt, sugar, and pH. Provided all the positive advantages of such IoUT farming tech-nologies, it is sensible for farmworkers to increasingly shift to agricultural drones and satellite systems for crop production 's development.

Drones allow farmers to monitor how much of the crop plants are during their various growing seasons. Farmworkers may also use drones to sprinkle undernourished crops with stimulants to allow them to return to daily existence. DroneFly estimates that perhaps the drones can disperse nitrogen 45 to 65 times more because they can get by hand.

12.7   CONCLUSION

IoUT agricultural software permits the gathering of relevant data by ranchers and farmers. Wide landowners and small farmers need to realize the IoUT market's opportunity for agriculture by implementing smart technology to improve their outputs' productivity and sustainability. With the increasingly increasing population, demand can be fulfilled effectively if the ranchers, as well as the small farmers, are effectively introducing agricultural IoUT solutions. Agriculture allowed by IoUT has continued to apply new technical approaches to the information tested over time. This helped bridge the gap between production and yielding quality and quantity. Data Consumed by collecting and storing information for real-time usage or storage from several sensors in a network assures rapid intervention and less damage to crops. With streamlined end-to-end smart operations and better implementation of business processes, inventory gets processed quicker and hits supermarkets in the shortest possible period.

KEYWORDS

- **device communication**
- **Internet of Things**
- **internet of underground things**
- **LoRaWAN**

REFERENCES

1. Manisha, B., & Ayush, K. A., (2020). War field spy and fighter robot. In: Harish, K., & Prashant, K. J., (eds.), *Recent Advances in Mechanical Engineering* (pp. 565–574). Singapore, Springer.
2. Yoo, S., Kim, J., Kim, T., Ahn, S., Sung, J., & Kim, D., (2007). A2S: Automated agriculture system based on WSN. In: *IEEE International Symposium on Consumer Electronics, 2007*. Irving, TX, USA.

3. Arampatzis, T., Lygeros, J., & Manesis, S., (2005). A survey of applications of wireless sensors and wireless sensor networks. In: *2005 IEEE International Symposium on Intelligent Control & 13th Mediterranean Conference on Control and Automation* (Vol. 1, 2, pp 719–724). Limassol, Cyprus.

4. Kotamaki, N., Thessler, S., Koskiaho, J., Hannukkala, A. O., Huitu, H., Huttula, T., Havento, J., & Jarvenpaa, M., (2009). Wireless in-situ sensor network for agriculture and water monitoring on a river basin scale in Southern Finland: Evaluation from a data users' perspective. *Sensors, 4*(9), 2862–2883. doi: 12.3390/s90402862.

5. Liu, H., Meng, Z., & Cui, S., (2007). A wireless sensor network prototype for environmental monitoring in greenhouses. *International Conference on Wireless Communications, Networking, and Mobile Computing (WiCom 2007)*. Shanghai, China.

6. Baker, N., (2005). ZigBee and Bluetooth-Strengths and weaknesses for industrial applications. *Comput. Control. Eng., 16,* 20–25.

7. IEEE, (2003). *Wireless Medium Access Control (MAC) and Physical Layer (PHY) Specifications for Low-Rate Wireless Personal Area Networks (LR-WPANs)*. In the Institute of Electrical and Electronics Engineers Inc.: New York, NY, USA.

8. Nandurkar, S. R., Thool, V. R., & Thool, R. C., (2014). Design and development of precision agriculture system using wireless sensor network. *IEEE International Conference on Automation, Control, Energy, and Systems (ACES).*

9. Joaquín, G., Villa-Medina, J. F., Nieto-Garibay, A., & Porta-Gándara, M. A., (2013). Automated irrigation system using a wireless sensor network and GPRS module. *IEEE Transactions on Instrumentation and Measurement,* pp. 0018–9456.

10. Vidya, D. V., & Meena, K. G., (2013). Real-time automation and monitoring system for modernized agriculture. *International Journal of Review and Research in Applied Sciences and Engineering (IJRRASE)* (Vol. 3, No. 1. pp. 7–12).

11. Kim, Y., Evans, R., & Iversen, W., (2008). Remote sensing and control of an irrigation system using a distributed wireless sensor network. *IEEE Transactions on Instrumentation and Measurement,* 1379–1387.

12. Wang, Q., Terzis, A., & Szalay, A., (2010). A novel soil measuring wireless sensor network. *IEEE Transactions on Instrumentation and Measurement,* 412–415.

13. Hayes, J., Crowley, K., & Diamond, D., (2005). Simultaneous web-based real-time temperature monitoring using multiple wireless sensor networks. *Sensors, 4.* IEEE.

14. Yoo, S., Kim, J., Kim, T., Ahn, S., Sung, J., & Kim, D., (2007). A2S: Automated agriculture system based on WSN. In: *IEEE International Symposium on Consumer Electronics, 2007.* Irving, TX, USA.

15. Arampatzis, T., Lygeros, J., & Manesis, S., (2005). A survey of applications of wireless sensors and wireless sensor networks. In: *2005 IEEE International Symposium on Intelligent Control & 13th Mediterranean Conference on Control and Automation* (pp. 719–724). Limassol, Cyprus.

16. Orazio, M., & Michele, B., (2011). A hybrid wired/wireless networking infrastructure for greenhouse management. *IEEE Transactions on Instrumentation and Measurement, 60*(2), 398–407.

17. Kotamaki, N., Thessler, S., Koskiaho, J., Hannukkala, A. O., Huitu, H., Huttula, T., Havento, J., & Jarvenpaa, M., (2009). Wireless in-situ sensor network for agriculture and water monitoring on a river basin scale in Southern Finland: Evaluation from a data users' perspective. *Sensors, 4*(9), 2862–2883. doi: 12.3390/s90402862.

18. Ayush, K. A., & Manisha, B., (2021). Analysis of offset quadrature amplitude modulation in FBMC for 5G mobile communication. In: Poonam, B., Meena, T., Valentina, E. B., & Rajeev, S., (eds.), *Proceedings of International Conference on Artificial Intelligence and Applications* (pp. 565–572). Singapore, Springer Singapore.
19. Ayush, K. A., & Manisha, B., (2020). Feature, technology, application, and challenges of the Internet of Things. In: Agrawal, R., Paprzycki, M., & Gupta, N., (eds.), *Big Data, IoT, and Machine Learning: Tools and Applications, Internet of Everything (IoE) Series* (pp. 255–276). Taylor & Francis Group.
20. Shafi, M., Andreas, F. M., Peter, J. S., Thomas, H., Peiying, Z., Silva, P. D., Fredrik, T., et al., (2017). 5G: A tutorial overview of standards, trials, challenges, deployment, and practice. *IEEE Journal on Selected Areas in Communications, 35*(6), 1201–1221.
21. Zhang, L., Pei, X., Adnan, Z., Atta, U. Q., & Rahim, T., (2017). FBMC system: An insight into doubly dispersive channel impact. *IEEE Transactions on Vehicular Technology, 66*(5), 3942–3956.
22. Bellanger, M., Le Ruyet, D., Roviras, D., Terré, M., Nossek, J., Baltar, L., Bai, Q., Waldhauser, D., et al., (2010). *FBMC Physical Layer: A Primer* (No. 4, pp. 7–12.). Phydyas.
23. Zhang, J., Minjian, Z., Jie, Z., Pei, X., & Tianhang, Y., (2017). Optimized index modulation for filter bank multicarrier system. *IET Communications, 11*(4), 459–467.
24. Zhang, L., Ayesha, I., Pei, X., Mehdi, M. M., & Rahim, T., (2018). Filtered OFDM systems, algorithms, and performance analysis for 5G and beyond. *IEEE Transactions on Communications, 66*(3), 1205–1218.
25. Kaur, S., Lavish, K., Gurjot, S. G., & Nuru, S., (2018). Survey of filter bank multicarrier (FBMC) as an efficient waveform for 5G. *International Journal of Pure and Applied Mathematics, 118*(7), 45–49.
26. Shaheen, I. A., Abdelhalim, Z., Fatma, N., & Reem, I., (2018). *Performance Evaluation of PAPR Reduction in FBMC System Using Nonlinear Companding Transform*. ICT Express.

CHAPTER 13

Smart Sensors for Digital Agriculture

HARMANDAR KAUR

Department of Engineering and Technology, GNDU RC Jalandhar Punjab–144007, India, E-mail: harmandargndu@gmail.com

13.1 INTRODUCTION

The ever-increasing population is stressing the natural resources and food that is one essential livelihood resource under severe stress. Compared with 2010, it is estimated that there will be a 25% rise in the world population by 2050, which is an addition of approximately 3 billion more people. This demand asks for a 25–70% increase in the global food production. This also increases the land requirement by 593 million hectares to meet the 25% increase in food demand [1, 2]. However, the reduction in the agricultural or fertile land cap available owing to the rapid depletion of the fertile top layer of soil and the depleting water resources are making this seem a farfetched goal. A whopping majority of 80% of the food is produced by more than 500 million small farms situated in the developing countries [3]. With the integration of smart sensors based IoT the precision agriculture technology can eliminate food damage and wastage. But the advances in technology are altering the agriculture practices, and the IoT sensor-based smart agriculture is helping to improve the farming efficiency and is a possible solution to meet the heightened demands in due time. Smart agriculture helps to monitor and enhance crop quality, increase productivity, manage the agriculture machinery, etc., with the aid of input from an individual sensor or a variety of sensing devices called

Harnessing the Internet of Things (IoT) for Hyper-Connected Smart World.
Indu Bala and Kiran Ahuja (Eds.)
© 2023 Apple Academic Press, Inc. Co-published with CRC Press (Taylor & Francis)

sensor system. In most of these domains, the sensors are required to be placed under the ground that calls for internet of underground things.

Smart sensors help farmers to supervise various aspects of precision agriculture even remotely. Better decision making relating to resources, crops, pesticides, etc., are possible that impact the environment. Smart agriculture dates back to 1980s with the use of GPS for mapping that helped to optimize the usage of water, weed control agents, fertilizers, and pesticides [4] and the advantages also include a huge environmental impact:

- Reduce fuel and energy consumption, bringing down the carbon dioxide emission levels;
- Reduce the fertilizer requirement or target area-specific fertilizer requirement, minimizes the release of nitrous oxide from soil;
- Reducing or precisely spraying pesticides;
- Monitoring and addressing the nutrient demands of the soil and the plant;
- Aiding in efficient water utilization.

Sensors are faculties of sight, sensing, and hearing at desired places in the field and for desired purpose to aid the farmer. These sensors act as supervisors for the farm practices and relay the collected data to the farmer or to the databases to obtain recommendations for the necessary action suitable for the given time. Hence, the sensors are indispensable for the IoT technology to function effectively as much as to the farmer. The rest of this chapter is arranged as follows: Section 1.2 explains the indispensable role of sensors in IoT and discusses the essential parameters for characterizing a sensor with suitable examples. In Section 1.3, the key drivers of technology in agriculture are presented and discussed, following which it also presents the different types of standalone sensors and sensor systems, respectively. These sections also elaborate the subdomain of agriculture wherein these sensors and sensor systems are highly suitable and aid in making the farm practices effective. The smart sensor technology has witnessed progress and a range of elaborate sensors are commercially available. In Section 1.4, an effort is made to present and discuss some of the commercially available sensors. The future roadmap of the smart sensors for agriculture is considered in Section 1.5 with useful insights explaining the futuristic research aspects of this field. This is followed by the conclusion.

13.2 SENSORS: INDISPENSABLE FOR IoT

IoT [5–7] enables the connection and communication of things/ objects which are connected to the internet with unique addresses. IoT has been defined as – "A dynamic global network infrastructure with self-configuring capabilities based on standard and interoperable communication protocols where physical and virtual 'things' have identities, physical attributes, and virtual personalities and use intelligent interfaces, and are seamlessly integrated into the information network, often communicate data associate with users and their environments" [8]. The data that is collected drives the rest of the system and is furnished by devices called sensors. A sensor is a gadget that identifies an adjustment in a physical boost that is transformed into a signal for interpretation or estimation. In general, it acquires a physical property, as is depicted in Figure 13.1, and converts it into a signal suitable for processing. It acts as an interface in doing so.

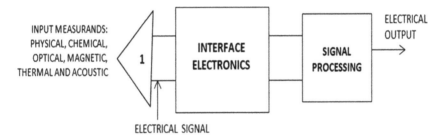

FIGURE 13.1   Sensor block diagram.

13.2.1   SENSOR PARAMETERS

The suitability of a sensor for a particular scenario can impact its performance widely. A number of features are essential to understand the sensor performance in order to obtain optimal values. The performance of a sensor is largely evaluated by the following parameters [9–11]:

1. **Calibration:** This is essential as the readings vary with time.
2. **Accuracy:** Enable statistical variation for measurement. It is the quantitative measure of vagueness or uncertainty observed value to an absolute value or in a general sense, how well the sensor is capable of measuring the environment in an absolute sense, e.g.,

a temperature sensor with 0.001°C accuracy is accurate within 0.001°C.

3. **Precision:** Describes the reproducibility of the measurement.

4. **Resolution:** The ability of a sensor to differentiate and recognize between small differences in readings. For example, a sensor with a resolution of 0.001°C, can read correctly up to 0.001°C.

5. **Range:** The range of values that can be read by the device. Precisely, it is the maximum and minimum value over which the sensor is operational and ceases to work properly or gets damaged below or above the range, e.g., a pressure sensor (100 m) deployed at 300 m depth.

6. **Repeatability:** The quality of a sensor to produce the same results when deployed for reading in the same environment. An inaccurate sensor can however possess repeatability.

7. **Drift:** It is the variation in a sensor with time or precisely electronic aging of sensor. Provisions in the form of calibration can be made to correct the drift. Drift is common in pressure sensors, especially for high pressure readings.

8. **Stability:** With a given input, you always get the same output. It is also the guaranteed accuracy for a said period of time.

9. **Sensitivity:** It is an absolute quantity, the smallest absolute amount of change that can be detected by a measurement.

10. **Hysteresis:** When output of a sensor lags the input generally corresponding to a linear change of the input, e.g., you get one curve on increasing pressure and another on decreasing. Many pressure sensors have this problem, for better ones, it can be ignored.

As stated already, sensors are an integral part of IoT which integrates processes, technologies, devices, and people to enable real-time decision making in communications, technical analytics, and other collaborations such as industry, automobiles, agriculture, etc., as is shown in the following Figure 13.2 [12]. The sensor is from where the technology intercepts and links to agriculture [13]. A typical IoT functional block has the sensor as an integral part of the device section as is shown in Figure 13.2. The sensing functionality is to extract the relevant data for the rest of the system to work on.

Broadly an IoT block comprises of the device, management, services, communication, security, and applications. The device block performs sensing, actuation, control, and monitoring. The data collected from the device section is forwarded for further data processing and analytics to the rest of the sections in the functional blocks of IoT. Here the sensors are essential to the device part of IoT functional block and their role in varied fields is different which is listed in Table 13.1.

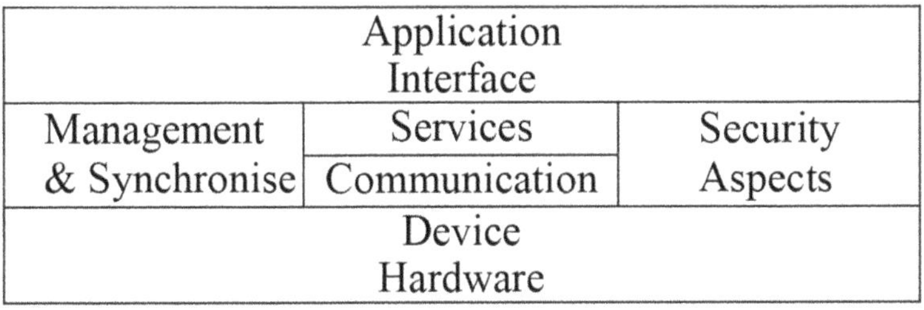

Application Interface		
Management & Synchronise	Services Communication	Security Aspects
Device Hardware		

FIGURE 13.2   IoT functional block.

TABLE 13.1   General Application Areas of IoT Sensors

Healthcare	Home Automation	Transport	Industrial	Environment
Light	Light	Gyroscope	Light	Light
Gyroscope	Gyroscope	Accelerometer	Temperature	Chemical
Contact	Contact	Temperature	Chemical	Pressure
Accelerometer	Accelerometer	Chemical	Pressure	Temperature
Chemical	Chemical	Magneto	Accelerometer	Humidity
Magneto	Magneto	Pressure	Hall effect	Magneto
Inertia	Temperature	–	–	Contact
Biosensors	–	–	–	–

IoT sensor technology is no exception to the agriculture domain, and is vital in making smart agriculture a reality, as it provides real time data for analyzing and super visioning the progression in various aspects of agriculture, hence ensuring timely decision making for better agriculture.

13.3   KEY DRIVERS OF TECHNOLOGY IN AGRICULTURE

The land suitable and available for cultivation is not going to increase due to incessant industrialization of urbanization [14] and rapid degradation of soil quality [15] especially the topsoil. However, the demand for agricultural products is ever-increasing both in food and its byproducts industry. Hence, even after the green revolution, there remains a demand-supply mismatch which might eventually broaden with time. Therefore, the integration of technology with agriculture can escalate the efficiency by improving the farm management reducing damage and wastage. The role of IoT sensor technology in agriculture is unparalleled and is bound to be an indispensable partner of agriculture in the times to come. Various aspects of agriculture (Figure 13.3) are already witnessing improvements with the intervention of IoT-based sensor systems. Broadly, the areas of crop yield, readiness to adapt to climate change, better resource (farm equipment, labor, vehicles, etc.), utilization, and management as well as automation are the key areas undergoing tremendous upgradation to enhance the production per unit area.

The key aspects of agriculture, as shown in Figure 13.3, are explained in subsections.

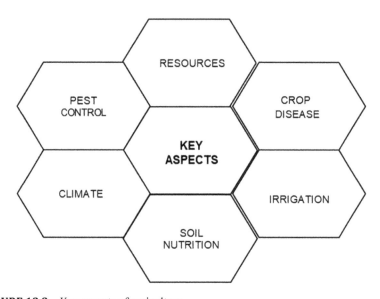

FIGURE 13.3   Key aspects of agriculture.

13.3.1   *SOIL HEALTH*

The soil quality is a measure of its nutrient content that includes soil type, fertilizer usage, cropping history, irrigation levels, etc. Soil moisture is integral to soil quality analysis. In most rural areas, this parameter is assessed via remote sensing, which also helps in acquiring data for drought like conditions given by the soil water deficit index. The soil mapping and field survey data for varied climate zones are also vital for prediction modeling. Soil salinity is essential and is tested by sensors set across the field that can map its variations over time.

13.3.2 IRRIGATION

Irrigation management that includes controlled irrigation with input water level inspection, controlled sprinkle, etc., is studied by crop water stress index (CWSI) [16, 17]. The CWSI for irrigation management needs temperature, satellite imaging, weather data and crop canopy information at different times. This generates specific irrigation index value for each required position in the field. Accordingly, the water usage can be optimized by the variable rate irrigation (VRI) by Crop-Metrics [18].

13.3.3 FERTILIZATION

The IoT sensor-based fertilizing methods help to apply variable rate technology (VRT) [19, 20] that optimizes the use of fertilizers based on the spatial patterns of nutrient requirements [21]. The spatial nutrient variation is studied with normalized difference vegetation index (NDVI) using satellite images [22] that are based on reflectance of near-infrared and visible light from vegetation to analyze vegetation vigor, density, health, etc. With fertigation [23] and chemigation [24] methods, the water-soluble materials like fertilizers and pesticides, etc., are applied through irrigation systems in a very precise manner [25] based on IoT sensor integration. GPS accuracy [26], autonomous vehicles [27], VRT, and geo-mapping [28] are key contributors to IoT-based smart fertilization. The variable rate fertilizer tools with input from yield maps and optical surveys of crop health that can be based on GPS. Accordingly, variable spraying is applicable wherein the time duration and the amount of spraying (fertilizer, pesticide, or herbicide, etc.), can be customized. These can be aid in weed elimination.

13.3.4   *CROP DISEASE, WEED, AND PEST MANAGEMENT*

IoT sensor systems with intelligent devices such as robots, drones, etc., are capable of real time pest management with the features of monitoring, forecasting damage, etc. [29, 30] compared to the conventional calendar based or prescription-based methods. The overall analysis for better crop yield as well as crop health by sensing, evaluating followed by the treatment. Field sensors are designed to acquire data such as plant health, environment monitoring, and pest monitoring for the crop cycle. Weed mapping can be managed by satellite imaging and GPS receivers working in collaboration and the maps hence obtained can be overlapped with the yield maps, spray, and field maps. Raw images for evaluation are acquired by the field sensors, remote sensing satellites, unmanned aerial vehicles (UAV). IoT-based automated traps [31, 32] can catch, count, and characterize the pests by uploading data to the cloud and analyzing it. The pest traps can allure, catch, and categorize the pest while notifying the farmer with the data of their count and the optimal amount of pesticide needed. This mechanism has saved 20 billion dollars in pest damage in the US alone.

13.3.5   *YIELD MAPPING AND MONITORING*

The analysis of yield quality and its maturity helps to determine the time for harvesting with the information of the weather taken together. This analysis covers various development stages of the crop over the period of the crop cycle. Yield quality is dependent on sufficient pollination to predict seed yield with the change of weather conditions [33, 34]. A yield monitor [35] installed on the harvester fetches real time harvest data to the farmer through the FarmRTX mobile phone application. Multiple optical sensors are deployed to monitor the health of the papaya plant yield [36].

The multiple domains in agriculture need to be attended with utmost attention and in a very precise manner. This can be extremely demanding for the farmer; however, there is a variety of devices that can aid the farmer in precise agriculture practices by supervising the domains for data collection and further analyzing the data to provide recommendations. In this manner, the precision agriculture aided by IoT-based sensors can help to improve the yield efficiency in a less burdensome manner for the farmer.

The sensors can be standalone type or heterogeneous sensor systems. The standalone sensors work by specialized mechanisms in order to perform the physical to electrical conversion of the parameter of interest. The working of various sensor variants that are essential to precise agriculture is explained in the following section with the domain of agriculture they are most suited.

13.4 STANDALONE SENSORS

Sensors more specifically wireless sensors are indispensable for smart agriculture as they play a vital role in collecting the data such as humidity, temperature, pest, crop yield, etc. The sensors can work independently, or different sensor types can form sensor systems, or even further wireless sensor network are available wherein various sensors can communicate amongst themselves.

13.4.1 ACOUSTIC SENSORS

These sensors work by detecting the variation of the noise levels upon interaction with materials of different kinds. Acoustic waves are mechanical pressure waves which are the principle in working of microphones and headphones that have a displacement transducer and diaphragm. The diaphragm's deflections are converted into an electrical signal. The operating frequency ranges from several hertz to several megahertz for the ultrasonic applications and even gigahertz in the surface acoustic-wave device [37]. Generally, they find application in areas such as seed classification [38] and pest monitoring and detection [39], etc. These sensors are economical and have good response times. However, at times they pick up stray noises such as that of the farm machinery which hampers temporarily the task assigned to it.

13.4.2 OPTICAL SENSORS

The optical sensor measures a physical quantity of light and converts it to a form that is readable by an integrated measuring device generally into an electrical signal. These can be either internal or external. External optical sensors are designed to collect and then transmit a defined light quantity, while internal sensors calibrate the directional variations in the light path. Optical sensors are capable of reading a wide variety of measurands by different optical mechanisms such as temperature, pressure, strain, velocity, displacement, vibrations, liquid level, pH value, force radiation, acoustic field, and electric field [40]. Soft water level sensing is used for assessing the water level or flow such as rain, stream, etc., with the provision of tunable time steps. For assessing soil moisture, soil density and composition in terms of its organic substances, clay, and mineral content, the reflection principle is used by these sensors [41, 42]. To be more precise, the working principle is based on the variations of wave reflections with respect to the electromagnetic spectrum. Optical fluorescence based sensors are used for measuring the plant and fruit maturation [43].

13.4.3   *OPTOELECTRONIC SENSORS*

Optoelectronics is the field of electronics with devices that source, detect, and control light. The light thus releases electrons from a semiconductor or metal surface, and, if there is an external electric field, the free charge carriers thus generated result in a photometric current proportional to the intensity of the light [44]. These sensors can distinguish between entities based on the reflection spectra and are used for mapping for instance, mapping the distribution of weed with the position [45, 46].

13.4.4   *ULTRASONIC RANGING SENSORS*

Ultrasonic distance/ranging sensors comprise of two membranes one that produces sound signal and the other detects its reflected echo. The distance of an object is determined from the time taken by the ultrasonic signal to reach back. The frequency in the ultrasonic wave, that is concentrated since sound at higher frequency dissipates less in the environment [47]. These sensors have a potential for use in varied applications owing to their adjustability, low cost, and ease of use. The sensor produces an ultrasonic signal for the purpose of ranging. The applications include object recognition, collision avoidance, uniform spraying, canopy detection, weed detection and yield coverage [48, 49].

13.4.5 TEMPERATURE SENSORS

Temperature Sensors are devised to measure heat energy of an object or system quantitatively. Contact type temperature sensors require physical contact with the object being sensed as they rely on the phenomena of conduction for temperature calibrations. These can detect solids, liquids, or gases over wide temperature ranges. However, non-contact temperature sensors work with convection and radiation mechanisms to monitor temperature variations and are used to detect liquids and gases [50]. These sensors utilize varied methods for temperature measurement and the output is produced by reading the variations. While some of these sensors depend on physical contact with the objects, others detect the radiant energy of the target object. They are vital for temperature measurements, in terms of weather forecasting, etc. Bee keeping project [51] has a temperature sensor which measures the temperature of the beehive and is powered by solar energy.

13.4.6 AIRFLOW SENSORS

This sensor determines the mass flow rate of air hence is also called a mass flow sensor. The temperature and pressure of the environment affects the air density. This sensor measures the volume of air but also compensates for its density. The sensor works based on the simple mechanism of the level of air pressure required for pushing a predefined volume of air to a specific depth. The variation is calibrated and hence providing an output. These sensors have been utilized for analyzing moisture values, soil properties, soil types [52]. These are used in general for yield quantity determination for more than two decades [53]. It works in the combine harvester part specifically of the tractor in combination with the other sensors such as storage unit, moisture sensor and a processor to generate a reliable output value.

13.4.7 LOCATION SENSORS

These sensors are aided by GPS satellites that are based on the triangulation principle to determine longitude, latitude, and altitude to within feet. Precise positioning is essential for achieving precision in agriculture. GPS integrated circuits such as NJR NJG1157PCD-TE1 are a good example of location sensors [54].

13.4.8   ELECTROCHEMICAL SENSORS

Electrochemical gas sensors determine the target gas concentration at an electrode by oxidizing or reducing target gas and measuring the resulting current. A negative or positive current flow is generated. The sensor section has three electrodes, i.e., the working, counter, and reference electrode [55]. They prove to be an economical alternate to the standard soil testing. They are used to analyze the soil features such as its content, pH, salinity [56].

13.4.9   ELECTROMAGNETIC SENSORS

These sensors are based on the GMR and TMR effect and are fabricated on ferromagnetic thin films. They are used in the direct magnetic field measurement and also in determination of distance, position, and rotation. These sensors are capable of near field electromagnetic measurements and assessing radiation characteristics [57]. These sensors depend on the electric circuits to measure the charge conduction or accumulation via contact or no contact. These work using electrical responses or transient responses. Essential nutrients such as nitrate content can be analyzed [58].

13.4.10   PRESSURE SENSORS

It finds application in measuring variables such as water level, fluid flow, speed, gas flow and altitude for gases or liquids. The sensors work by assessing the compaction with the help of a pressure sensor. The change in the compaction gives the output [59]. For instance, the sensor can be deployed to analyze by cutting through the soil and recording the force required by the strain gauge.

13.4.11   EDDY COVARIANCE SENSORS

The eddy covariance method is a vital atmospheric measurement technique to determine vertical turbulent fluxes for given atmospheric boundary layers. It estimates heat, momentum, water vapor and gas fluxes typically carbon dioxide and methane. These utilize the surface atmospheric energy flux values for determining the exchange of water vapors, methane, carbon dioxide, oxygen, or other gases. The close chamber method could be used [60]. However, these sensors are preferred owing to continuous flux measuring capability with a greater precision.

13.4.12 *LIDAR SENSORS*

Light detection and ranging (LiDAR) are remote sensing system that measures vegetation height across wide areas and can calculate the distance to a target by using laser light followed by the determination of the reflected light with a sensor. 3D laser scanning has 3D scanning and laser scanning combined. High-resolution maps are obtained using lidar technology. This sensing technique has a wide set of applications within the agriculture domain with a prominent feature of 3D mapping with the aid of GPS [61]. It has promised to be an important tool for finding erosion area, forecasting crop production, estimating biomass, finding the soil characteristics [62].

13.4.13 *NANOSENSORS*

1. **Carbon Nano Tube:** The carbon molecule flexible cylindrical tube is held by van der Waals with varied number of wall construction. These nanotubes have been used to supplement the soil with iron and have increased the crop yield [63]. This has increased the growth of the plant and its root system for wheat, garlic, maize, peanut, etc. [64]. Fullerene a variant of carbon nanotube helps in enhancing tomato production by retaining the water content [65]. These nanotubes enhance nutrient content, water retention hence the growth of the plant.
2. **Nano Aptamers:** These single-strand nucleic acid having nano or picomolar ranges and are good for high affinity target-specific binding [66]. Systematic evolution of ligands by exponential enrichment [67] is used for ligand specific binding. These aptamers are efficient disease detectors, increase the crop yield and resistance. The aptamer sensors are vital for determining the toxicity level in food crops sprayed with herbicide and pesticide [68]. Insulin binding aptamers are used for detecting light emission from a cell labeled with photoluminescence.

3. **Smart Dust Technology:** These are micro-sized electrochemical almost undetectable sensors monitored via wireless radios irrespective of their location [69]. These have been deployed for sensing, processing, and computing parameters related to environmental hazards, energy usage, temperature, traffic tracking, etc. They are capable of detecting almost everything in their vicinity.

13.4.14   SMART PHONE SENSOR

Smartphone-based sensors that being used in various agriculture applications:

1. **Image Sensor (Camera):** The images obtained from this sensor can be readily used for leaf area indexing, soil erosion study, harvest quality, disease detection, fruit maturation, chlorophyll content [70–72].
2. **GPS Sensor:** Provides the location details of the entity to which it is attached. It is largely used for machine tracking, crop mapping and land management [73, 74].
3. **Microphone:** The microphone in the mobile phone is analyzed for its use in machine maintenance and bug detection [75, 76].
4. **Gyroscope:** Senses the angular velocity and has been studied for canopy measurement and equipment motion [77, 78].
5. **Accelerometer:** Senses the acceleration of the entity under observation, such as farmworkers and machinery [71, 74].
6. **Barometer:** Measures altitude and is used in vertical agriculture for measuring the elevation [80]. Inertial sensor: it is used to determine the precise distance of seed, plant from any other object [81].

13.4.15   SMARTPHONE APPLICATIONS

Smartphone applications have evolved with the features like data aggregation, speedy processing and incorporate IoT ideals that collect data from sensors, analyze it and give recommendations based on the real time information. The data collected through the sensors in the phone or in the field is accessible by the application and which gives valuable insights related to seeding, weeding, fertilizing, and watering, etc., to farmers. Several applications have been developed to aid farmer in timely decision making and hence improving farming efficiency:

1. **Water Requirement Analysis:** Determining water requirement from photos and brightness through leaf area indexing.
2. **Crop Disease Detection, Analysis, and Diagnosis:** Experts analyze the photos of suspect plants.
3. **Soil Quality Assessment:** Using soil images, pH, and chemical data from sensors.
4. **Crop Harvesting Timelines:** Ripeness prediction by camera photos with UV and white lights.
5. **Fertilization Estimation:** Nutrient deficiency is evaluated by soil sensors and leaf color sensor.

When specialized applications improve farm productivity by analyzing soil, crop, weed, and pest variables, as well as offer valuable feedback for agricultural decisions, the small farmer's quality of life can noticeably improve [82]. The individual standalone sensors are very appropriate for the specific data collection. However, it is also feasible to collect the data for a specific purpose, for instance, for soil moisture content from multiple sensors such as environment humidity sensor, soil moisture content and use the data for further decision making. This type of heterogeneous sensor usage is called a sensor system. In another approach, the sensor systems can be used for collecting data of varied aspects also. In the following section, the sensor systems and their role in different agriculture aspects is discussed along with the details of the commercially available sensor systems.

13.5 SENSOR SYSTEM

As discussed earlier, sensor systems are an accumulation of more than one sensor, working together for accomplishing a given sensing task. The sensor system can be used for detecting one or more parameters of importance. The sensor systems prove helpful as they collect data on to a single system reducing latency and improving processing capabilities while ensuring better resource utilization.

13.5.1 TELEMATICS SENSORS

Telematics is a method of monitoring an asset (tractor, etc.), via the connected hardware or sensors. These sensors provide data from the remote locations to avoid visiting the same place [83]. These have been deployed as anti-theft sensors [84]. They yield information of the functioning of the farm components.

Telemetry data can be relayed via a satellite by the automatic packet reporting system [85]. The farmers can remotely access the information of their farm machinery and vehicles in real-time for any maintenance or damage. This information can be relayed to a mechanic so that necessary action can be taken [86]. In Ref. [87], various soil properties relating to land degradation are estimated by resolution imaging spectroradiometer (MODIS) sensor.

13.5.2   REMOTE SENSING

Remote sensors determine the energy reflected by Earth to obtain input as these are mounted on airplanes or satellites. The data from ground sensors and that from the satellite sensor can be added up to get a better understanding of the situation. The sensors of this type obtain, manipulate, analyze, and manage data pertaining to geography. These also hold a wide scope of applicability in agriculture as they are significant in acquiring land cover and degradation map, plant or pest identification, crop assessment and forecast of production [88]. The global coverage via satellites helps to acquire, process, and distribute the data [89, 90].

13.5.3   DIELECTRIC SENSOR

Dielectric soil moisture sensors [91] work with dielectric constant that changes with the change in property of the medium under test for instance, the medium is soil, and the property can be moisture. When used together with the rain gauge sensor, these can aid to analyze the soil moisture content better.

13.5.4   FPGA SENSOR

The speed and power consumption are enhanced by the optimized field-programmable gate arrays (FPGAs) compared to microcontrollers used in sensors. FPGAs can provide sensors with processing capabilities as per

IEEE 1451. These can include running calibrations, configuration, and self-tests as per the requirement. The same interface can be accessed for a group of several sensors working together as an array and operational through the wireless forms of communication [92, 93]. The use of these sensors is in initial stages specifically for agricultural applications as these are reconfigurable. These have been used for humidity, irrigation, and transpiration measurements [90, 93]. However, a major concern is more power consumption that renders these unsuitable for real time monitoring [94]. Optical sensors together with the microwave scattering they can characterize canopies [95]. Sensors and vision devices are used in designing Agribot [96] that can sow seeds accurately by analyzing the right distance and the right depth of sowing. The vision devices or components can be other than a camera, for instance, in Ref. [97] the LEDs with infrared, radiation receivers, laser-LEDs are used for sowing. The output is a voltage variation associated with the seed movement through the sensor.

13.5.5 BEAN-IoT SENSOR

The sensors can go beyond the farm area, and a grain sized plastic sensor called BeanIoT [98] accompanies the grains to the silos and determines the storage conditions related data that is sent to the farmer. This sensor system shown in Figure 13.4 has a humidity sensor, gyroscope, electronic compass, and Bluetooth and can determine moisture, temperature, air quality, gas levels, altitude, and the movement of the grain inside the silo. Bean IoT sensors can detect unusual values and this alert to a smartphone. Remaining time these sensors remain in sleep mode for saving energy. They can charge wirelessly and have a battery life of 14 months.

13.5.6 UAV

The drones or unmanned aerial vehicles as shown in Figure 13.5 are revolutionizing agriculture as they can be used individually or in swarms, groups of drones equipped with a variety of sensors, including 3D cameras, can work together to provide farmers management capability for their land with the vision elevated in the sky. These are effective in rough terrains and can aid in precision agriculture as the spraying of pesticides, fertilizers, etc., can be target areas [99, 100].

FIGURE 13.4   Bean IoT sensor system nodes.
Source: Ref. [98].

FIGURE 13.5   IoT-enabled sensor system supported UAV drones for agriculture.
Source: Ref. [100].

13.5.7   GREEN SENSORS

The creation of sustainable energy efficient IoT involves the use of green sensors that have low energy demand. Green IoT has been defined as [101]: "The energy-efficient procedures (hardware or software) adopted by IoT either to facilitate reducing the greenhouse effect of existing applications and services or to reduce the impact of greenhouse effect of IoT itself. In the earlier case, the use of IoT will help reduce the greenhouse effect, whereas in the next case further optimization of IoT greenhouse footprint will be taken care of. The entire life cycle of green IoT should focus on green design, green production, green utilization and finally green disposal/recycling to have no or very small impact on the environment." In green WSNs the following are ensured for energy efficiency [102–105]:

1. **Sleep Mode:** Sensor nodes only work when required. This reduces the energy consumption per node, and for a sensor network comprising of tens or hundreds or more nodes huge power saving over long periods of operation are achievable [106].

2. **Energy Harvesting:** Generate power from the environment, e.g., solar energy, kinetic energy biomass, oxygen, wind, etc. Generally, the renewable resources are green power sources that can replenish. For instance, solar energy can increase the lifetime of a WSN to ideally infinite value, which is a boon for smart agriculture.

3. **Radio Optimization Techniques:** Modulation optimization, transmission power control, directional antennas, cooperative communication, and energy-efficient cognitive radio [107] can be used. The cognitive radio is aware of its environment and can change its operation mode by varying hardware and/or software modes.

4. **Data Reducation Mechanisms:** Adaptive sampling, aggregation, compression, network coding is used.

5. Large multimedia files require a lot of energy for transmission. If it is not required or can be substituted by simpler files energy saving are possible. User behavior assessment can be utilized for predicting the data delivery needs of the user.

6. **Energy-Efficient Routing Techniques:** Energy as a routing metric, cluster architectures, multipath routing, node mobility and relay node placement are needed. Efficient routing methods can provide optimal path in terms of minimum energy route that are generally shorter routes [108]. For obtaining short paths, the network mechanisms relating to the routing needs can be applied [109]. Energy-saving architectures [110], compressive sensing [111] and cooperative relaying [112, 113] have been used.

Rapid advances in agriculture-technology integration are proving vital, and there are many sensor systems designed and commercialized to help the farmer for a better yield quantity and quality. Further improvement in the sensor system requires an insight into the already available systems and an understanding of their constitution and their role in the agricultural subdomains. The following section creates an understanding of the different commercially available sensor systems outlining their importance in improving agricultural practices.

13.6   COMMERCIALLY AVAILABLE SENSOR SYSTEMS

Agriculture sector is witnessing a boost thanks to the improved sensor systems. A number of commercial sensor systems are aiding the farmer in data collection, supervision, and proper recommendations and guidance. These smart sensors are connected to end-user or databases through the bridge layer that can have any of the following technologies to its aid for instance Bluetooth, Wi-Max, GSM, satellite, RFID, Zigbee, Wi-Fi enabled, with Lora, Sigfox or cellular communication which is followed by the sensing layer. The sensing layer comprises of standalone sensors which are already discussed or sensor systems. The commercially available sensor systems have enhanced over the years and are proving vital in improving the agriculture efficiency. Some of the sensor systems commercially available are discussed with a focus on their role in the agriculture sub-domain for which these are designed.

1. **LOUP 8000i:** In this sensor system, the crop yield sensing is accomplished by optical sensors and impact plate technologies to provide adequate results, with the Loup Electronics Elite Yield Monitor for display. The optical sensor has no effect on the crops moisture content; hence results are obtained in a single calibration, unlike the other yield systems. Field mapping feature is also included [114] (Figure 13.6(a)).

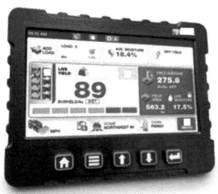

Optical Sensor
Impact Plate Technology
Field Mapping
Build-in Display Console

FIGURE 13.6(a)   LOUP 8000i.
Source: Ref. [114].

2. **XH-M214:** This sensor system has a digital display for soil moisture calibrated by the electrode in the soil. It is capable of measuring the moisture content of the soil and can input the water to the soil if the level falls below the required predefined value. Once the prescribed water content is reached, the water pump is turned off [115] (Figure 13.6(b)).

Soil moisture sensor
Build in led display
Controls input water supply level

FIGURE 13.6(b)   XH-M214.
Source: Ref. [115].

3. **Ag Premium Weather:** This offers intelligent field-level precision with up-to-date precipitation tracking and temperature assessing. The current and historical precipitation data is stored and is available anytime for better decision making including yield potential analysis, yield quantity prediction and operational planning [116] (Figure 13.6(c)).

Intelligent field-level precision
Temperature assessing
Up-to-date precipitation tracking
Yield quantity prediction

FIGURE 13.6(c)   Ag premium weather.
Source: Ref. [116].

4. **FI-MM:** This sensor system is used to monitor the growth rate of fruits with the displacement sensors in a range of 7 to 160 mm. It comprises of an LVDT transducer with DC powered signal conditioner and a holding clip. The internal water content and the growth rate can be analyzed. Generally, the growth rate is affected by a number of factors such as water content, light exposure, etc., in absence of which the fruit shrinkage can occur. This sensor system finds application in monitoring the fruit condition in storage [117] (Figure 13.6(d)).

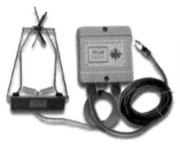

monitor growth rate of fruits
internal water content
displacement sensors
LVDT transducer

FIGURE 13.6(d)   FI-MM.
Source: Ref. [117].

5. **MP406:** The MP406 ensures high accuracy results for soil moisture content monitoring since the outcomes are independent of the soil salinity or the temperature of the surroundings. The sensor system can take measurements continuously or through temporary burial through connection to a soil moisture meter [118] (Figure 13.6(e)).

for soil moisture content
monitoring
independent of the soil
salinity or the temperature

FIGURE 13.6(e)   MP406.
Source: Ref. [118].

6. **DEERE2630:** The farm vehicle comes equipped with multiple sensor systems to ensure fatigue less farm practices and reduced stress for the farmer. The sensors enable automated truck and

trailer filling enabling harvest anytime and in turn improving the overall harvest and hence farm efficiency [119] (Figure 13.6(f)).

Harvest anytime
Multiple sensor systems
Automated truck and trailer
filling

FIGURE 13.6(f) DEERE2630.
Source: Ref. [119].

7. **SOL CHIP COM:** This sensor system comes with the advantage of enabling maintenance free and autonomous ultra-compact solar-powered wireless tag and operates for years. SCC-S433 has the capability to power, control, and wirelessly connect a variety of sensors to cloud [120] (Figure 13.6(g)).

autonomous ultra-compact
solar-powered
power, control and wirelessly
connect variety of sensors to
cloud

FIGURE 13.6(g) SOL CHIP COM.
Source: Ref. [120].

8. **DEX70:** It has a full bridge strain gauge with a flexible arm and is caliper styled. This series has highly precise electronic dendrometers that determine the growth of plant fruits and stem. The stem size variations and water balance of plants are sensitive to environmental conditions over time are easily monitored with a temperature compensated dendrometer. The environment

conditions may include conditions of elevated ozone, water stress and other atmospheric pollutants. Since it can operate in long term mode it gives output of diurnal and long term growth of the plant in millivolt range. Applications for screening plants for growth rate and stress tolerance are also common [121] (Figure 13.6(h)).

screening plants growth
rate and stress tolerance
electronic dendrometers

FIGURE 13.6(h)   DEX70.
Source: Ref. [121].

9. **CI-340:** This is a portable sensor system mounted with multiple sensor types to aid in the assessment and determination of respiration, transpiration, photosynthesis, stomatal conductance, PAR, and internal CO_2.

    The modules can help control H_2O, light intensity, temperature, CO_2 and find chlorophyll fluorescence. The chambers allow any leaf size to be tested. Further, to minimize the measurement delay direct chambers can be used for connection to water or carbon dioxide analyzer [122] (Figure 13.6(i)).

Portable sensor system
Multiple sensor types :
respiration, transpiration,
photosynthesis, stomatal
conductance, PAR, CO_2

FIGURE 13.6(i)   CI-340.
Source: Ref. [122].

10. **Wind Sentry 03002:** The sensor system determines the speed and direction of wind using a three-cup anemometer and a wind vane. Without signal conditioning requirements it can send data to the device support data loggers [123] (Figure 13.6(j)).

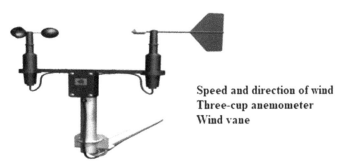

Speed and direction of wind
Three-cup anemometer
Wind vane

FIGURE 13.6(j)   Wind sentry 03002.
Source: Ref. [123].

11. **AQM-65:** This sensor system is a highly precise air monitoring system with the capacity to differentiate and measure up to 20 gaseous or particulate pollutants and environmental parameters simultaneously. The pollutants including carbon monoxide, carbon dioxide, ozone, sulfur dioxide, nitrogen dioxide, nitrogen oxides, volatile organic compounds, hydrogen sulfide, particulate matter, noise, and meteorological parameters such as temperature, rainfall, humidity, pressure, wind speed and direction [124] (Figure 13.6(k)).

Air monitoring system
Meteorological parameters
Measure up to twenty gaseous or
particulate pollutants

FIGURE 13.6(k)   AQM-65.
Source: Ref. [124].

12. **POGO-PORTable:** This is a soil sensor can continuously monitor moisture, salinity, and temperature of soil with good accuracy [125] (Figure 13.6(l)).

soil sensor
monitor moisture, salinity
and temperature

FIGURE 13.6(l)   POGO-PORTable.
Source: Ref. [125].

13. **SF-4/5:** This is a sap flow sensor system are based on modified thermal dissipation method which helps in sap flow measurement that is present generally in plant branches, stems, and roots and it measured in sapwood part of the xylem of the plant [126] (Figure 13.6(m)).

Sap flow sensor system
Sap flow measurement
Modified thermal dissipation
method

FIGURE 13.6(m)   POGO-PORTable.
Source: Ref. [126].

14. **MET STATION ONE:** This sensor system measures wind direction, wind speed, relative humidity, barometric pressure, and ambient temperature and has optional sensors included for rain and solar radiation determination [127] (Figure 13.6(n)).

Sensor system
has optional sensors
wind direction, wind
speed, relative humidity,
barometric pressure

FIGURE 13.6(n)   MET STATION ONE.
Source: Ref. [127].

15. **SD-6P:** SD-type sensor system is a highly precise incremental LVDT-based sensor that monitors micro-variations of stem diameter in micron range. Plant stem diameter is altered by water intake and plant growth. The daily contraction amplitude and the daily maxima trend are used for assessing the normal plant growth rate [128] (Figure 13.6(o)).

incremental LVDT-based
sensor
monitors micro-variations
of stem diameter
for assessing the normal
plant growth

FIGURE 13.6(o)   SD-6P.
Source: Ref. [128].

16. **YieldTrakk:** This sensor system relies on non-contact optical sensors to determine live yield data and is installed in the clean grain elevator. It is equipped with an electronic control unit which converts the measurements to a weight value corresponding to the crop quantity harvested. Another set of sensors is deployed

that measure crop moisture content and include it to the crop yield output data. To obtain accurate results, the measured slope variations are used for correcting the signal for any changes in machine angle [129]. A multitude of other sensor systems are available in the market, some of these are briefly discussed with their area of application in agriculture and the sensor variants used (Figure 13.6(p)).

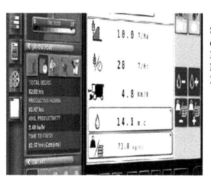

FIGURE 13.6(p)   YieldTrakk.
Source: Ref. [129].

- Biponics [130] automates home planting using a sensor system comprising of humidity, pH, brightness, temperature, etc., and the web services to gather the information on the hydroponic system requirements.
- Botanicalls [131] uses a sensor system to analyze the plant needs such as inadequate plant hydration, etc., and conveys the information to the user.
- Edyn [132] is a cloud enabled garden monitoring sensor system equipped with temperature, moisture, light, soil, humidity, acidity, light sensors which is solar-powered. The user can decide the action depending on the analysis of the data procured from the sensor system.
- Parrot [133] is a Bluetooth enabled sensor system that assesses the data relating to soil salinity, temperature, humidity, fertilizer from its vicinity and aids in decisions related to the plants health.
- Plantlink [134] is a soil hydrating sensor system with cloud support that adapts to the temperature of its surroundings by acquiring data relating to it. It also has a supporting database to help it adapt

the plant moisturizing levels depending on the nature of the plant species.

- Harvestbot [135] is a sensor system with actuators and cloud server support. It is useful in large open areas and comprises of soil, humidity, temperature, acidity, light, CO_2 ppm level sensors. The data collected by these devices can be vital for the health assessment of the plants.
- Iro or Rachio [136] is a Wi-Fi enabled irrigation sprinkler controller operational globally which easily adjusts to the local weather conditions.
- Spruce [137] is an irrigation controller with soil moisture and temperature sensors to its aid with global data access.
- Koubachi [138] is an irrigation sprinkler controller specifically designed to cater to garden area with its array of sensors including ambient light, soil moisture, soil temperature and air temperature whose number can be customized by the user.
- Niwa [139] is an indoor hydroponic system is Wi-Fi enabled to alert user about the requirement of the plant with light, ventilation, humidity, water level, and heat sensors.

Table 13.2 summarizes the sensors and the sensor systems discussed and deployed in various significant sub-areas of agricultural practices such as soil nutrition, irrigation, fertilization, crop disease, yield monitoring, and weather changes.

The discussion on the commercially available sensors provides insights and can help to develop innovative approaches for further improvements in the smart sensors. The futuristic approaches in this ever evolving area of smart sensors technology are discussed next and hold a lot of promise for the enhancement of technology aided agriculture practices.

13.7   FUTURE ROADMAP

The IoT sensor aided smart agriculture is providing promising outcomes in all aspects of agriculture. The integration of precision agriculture with technology is improving the agricultural outcomes, reducing wastage, and minimizing detrimental effects on the environment. This integration has a long way to go as it is still evolving and enhancing to meet the future world food demands. Below a few possible future trends are discussed that may contribute to the futuristic smart agriculture:

TABLE 13.2   Classification of Application Subdomain Specific Sensors

Agriculture Aspects	Standalone Sensors	Sensor Systems
Soil nutrition	Optical sensor [40–42], airflow sensors [52], electrochemical sensors [56], electromagnetic sensor [58], lidar sensors [62], nanosensor [63], image sensor [70–72]	Remote sensing [88], Dielectric sensor [91], xh-m214 [115], mp406 [118]. POGO-portable [125], Biponics [130], EDYN [132], Parrot [133], Harvest bot [135]
Irrigation	Pressure sensor [59], nanosensor [65], image sensor [70–72]	Remote sensing [88], FPGA sensor [90, 94], UAV [99, 100], sf-4/5 [126], SD-6P [128], Biponics [130], Botanicalls [131], EDYN [132], Plantlink [134], Harvest Bot [135], Rachio [136], Spruce [137], Koubachi [138], NIWA [139]
Fertilization	Nanoaptamers [68], image sensor [70–72]	Remote sensing [88], UAV [99, 100], Parrot [133]
Crop disease, pest control, weed mapping	Acoustic sensors [39], optoelectronic sensors [45, 46], ultrasonic ranging sensors [48, 49], nanoaptamers [68], image sensor [70–72], microphone [75, 76]	Remote sensing [88], UAV [99, 100], Harvest bot [135]
Yield, monitoring, and mapping	Acoustic sensors [38], optical sensors [43], ultrasonic ranging sensors [48, 49], airflow sensors [52], location sensor [54], lidar sensors [62], nanosensor [63], image sensor [70–72]	Remote sensing [88], FPGA sensor [95], Agribot [96], FPGA vision sensors [97], Beaniot [98], UAV [99, 100], Loup 8000i [114], FI-MM [117], dex70 [121], ci-340 [122], SD-6P [128], yieldtrakk [129]
Resources	Ultrasonic ranging sensors [48, 49], location sensor [54], accelerometer [71, 74], image sensor [70–72], microphone [75, 76], gyroscope [77, 78], inertial sensor [81]	Telematics sensor [83], UAV [99, 100], deere2630 [119]
Climate	Temperature sensors [51], eddy covariance sensors [60]	Fpga sensor [90, 94], AG premium weather [116], Wind sentry 03002 [123], aqm-65 [124], Met station one [127], Biponics [130], EDYN [132], parrot [133], Plantlink [134], Harvest bot [135], Spruce [137], Koubachi [138], NIWA [139]

- More than 75 million IoT-enabled sensor aided devices are already deployed in agriculture industry and more are on their way. It is predicted that as of 2050 the smart agriculture alone would amount to around 4.1 million data points on a daily basis [140]. For instance, in wireless sensor network [141] that are equipped with a number of sensor nodes and a base station or sink node. Sensor nodes are for gathering the requisite data from their surroundings and communicate it to the base node for further processing. WSNs are based on the IEEE 802.13.4 standard [142].
- Furthermore, precision agriculture with the aid of varied types of sensors is providing solutions to the global food demand by collecting data, adapting to changing environment and with proper utilization of resources. These technologies contribute to reduced pollution, global warming, and hence conservation. The future precision agriculture is set to witness data acquisition from smaller and smarter sensors, increased usage of autonomous vehicles, etc. [143].
- The sensor-cloud computing [143] integration has great potential and a vital role for enhancing the smart agriculture in the near future. It has the capability to improve data access and device management by filtering the necessary information from the available data. Similarly, a mobile on the go sensor-cloud platform can be designed.
- Vision and machine learning enables robots to train using their surroundings with the aid of smart sensors. However, the agriculture robotics is still in its early stages since it needs to undergo a lot of R&D trials [144].
- A lot in the fate of smart agriculture in the near future depends on the evolution of combinational sensors and sensor fusions. The necessary crop data furnished will impact yield quantity and quality to keep pace with the increasing demand for food [145].
- IoT smart sensors implementation based solutions could be a great opportunity for the overall agricultural landscape, including seeding stage to selling stage and beyond [146, 147].

13.8   CONCLUSION

IoT and sensor based agriculture is aimed at enhanced crop production owing to the availability of real time data for decision making, thus avoiding damage and wastage. The vast variety of sensors aid the farmer in covering a huge range of potential parameters such as soil quality, seed count, pest control, farm machinery maintenance, disease detection, moisture level analysis, water supply management, 3D mapping, weed control, terrain management, crop price updates to until the crop reaches the market. The sensors have gained from the advancements in the image and signal processing domains, especially for crop health and quality. All this is imperative to make agriculture sustainable. The integration of IoT and agriculture has revolutionized the farmers vision by making them aware of the requirements at every stage and also providing insights.

KEYWORDS

- **digital agriculture**
- **smart agriculture**
- **sensor**
- **IoT**

REFERENCES

1. *World Population Projected to Reach 9.8 Billion in 2050, and 11.2 Billion in 2100.* Available: https://www.un.org/development/desa/en/news/population/world-population-prospects-2017.html (accessed on 16 November 2021).
2. Hunter, M. C., Smith, R. G., Schipanski, M. E., Atwood, L. W., & Mortensen, D. A., (2017). Agriculture in 2050: Recalibrating Targets for Sustainable Intensification, *BioScience, 67,* 386–391.
3. *Food Production Must Double by 2050 to Meet Demand from World's Growing Population.* Available: https://www.un.org/press/en/2009/gaef3242.doc.htm (accessed on 16 November 2021).
4. www.agritechtomorrow.com/article/2019/02/top-article-for-2019-smart-sensors-in-farming/11247 (accessed on 16 November 2021).
5. Atzori, L., Iera, A., & Morabito, G., (2010). The Internet of Things: A survey. *Comput. Netw., 54,* 2787–2805. doi: 10.1016/j.comnet.2010.05.010.
6. Gubbi, J., Buyya, R., Marusic, S., & Palaniswami, M., (2013). Internet of Things (IoT): A vision, architectural elements, and future directions. *Future Gener. Comput. Syst., 29,* 1645–1660. doi: 10.1016/j.future.2013.01.010.
7. Lin, J., Yu, W., Zhang, N., Yang, X., Zhang, H., & Zhao, W., (2017). A survey on Internet of Things: Architecture, enabling technologies, security, and privacy, and applications. *IEEE Internet Things J., 4,* 1125–1142.

8. Kranenburg, R. V., (2008). *The Internet of Things: A Critique of Ambient Technology and the All-Seeing Network of RFID*. Institute of Network Cultures.
9. Coombs, & Clyde F., Jr., (2000). *Electronic Instrument Handbook* (3rd edn.). New York: McGraw-Hill. https://www.accessengineeringlibrary.com/content/book/9780070126183 (accessed on 16 November 2021).
10. https://www.mccdaq.com/TechTips/TechTip-1.aspx (accessed on 16 November 2021).
11. https://ocw.mit.edu/courses/mechanical-engineering/2-693-principles-of-oceanographic-instrument-systems-sensors-and-measurements-13-998-spring-2004/readings/lec2_irish.pdf (accessed on 16 November 2021).
12. Chien, L. J., Drieberg, M., & Sebastian, P., (2016). A simple solar energy harvester for wireless sensor networks. *IEEE International Conference on Intelligent and Advanced Systems (ICIAS)*.
13. Elijah, O., Rahman, T. A., Orikumhi, I., Leow, C. Y., & Hindia, M. N., (2018). An overview of Internet of Things (IoT) and data analytics in agriculture: Benefits and challenges. *IEEE Internet Things J., 5*, 3758–3773.
14. 68% of the World Population Projected to Live in Urban Areas by 2050, Says UN. Available: https://www.un.org/development/desa/en/news/population/2018-revision-of-world-urbanization-prospects.html (accessed on 16 November 2021).
15. *Why Soil is Disappearing from Farms*. www.bbc.com (accessed on 16 November 2021).
16. Zhang, L., Dabipi, I. K., & Brown, W. L., (2018). Internet of Things applications for agriculture. In: Hassan, Q., (ed.), *Internet of Things A to Z: Technologies and Applications*.
17. *Irrigation & Water Use*. https://www.ers.usda.gov/topics/farm-practices-management/irrigationwater-use/ (accessed on 16 November 2021).
18. LaRue, J., & Fredrick, C., (2012). Decision process for the application of variable rate irrigation, *Amer. Soc. Agricult. Biol. Eng.* Dallas, TX, USA, Tech. Rep.
19. Colaço, A. F., & Molin, J. P., (2017). Variable-rate fertilization in citrus: A long term study, *Precis. Agricult., 18*, 169–191.
20. Basso, B., Dumont, B., Cammarano, D., Pezzuolo, A., Marinello, F., & Sartori, L., (2016). Environmental and economic benefits of variable rate nitrogen fertilization in a nitrate vulnerable zone. *Sci. Total Environ., 545, 546*, 227–235.
21. *Why IoT is Reinventing Plant Fertilization*, (2017). Available: https://www.iof2020.eu/latest/news/2017/09/why-the-Internet of Things-is-reinventing-plant-fertilization (accessed on 16 November 2021).
22. Benincasa, P., Antognelli, S., Brunetti, L., Fabbri, C., Natale, A., & Sartoretti, V. V. M., (2018). Reliability of NDVI derived by high resolution satellite and UAV compared to in-field methods for the evaluation of early crop N status and grain yield in wheat. *Exp. Agricult., 54*, 604–622.
23. Raut, R., Varma, H., Mulla, C., & Pawar, V. R., (2017). Soil monitoring, fertigation, and irrigation system using IoT for agricultural application. *Intelligent Communication and Computational Technologies* (pp. 67–73). Singapore: Springer.
24. González-Briones, A., Castellanos-Garzón, J. A., Martín, Y. M., Prieto, J., & Corchado, J. M., (2018). A framework for knowledge discovery from wireless sensor networks in rural environments: A crop irrigation systems case study. *Wireless Commun. Mobile Comput., 2018*, 6089280.
25. Palomino, G., & Miguel, J., (2017). Protected crops in SPAIN: Technology of fertigation control. *Proc. Agri-Leadership Summit*. Haryana, India.

26. Shi, J., Yuan, X., Cai, Y., & Wang, G., (2017). GPS real-time precise point positioning for aerial triangulation. *GPS Solutions, 21*, 405–414. doi: 10.1007/s10291-016-0532-2.

27. Khan, N., Medlock, G., Graves, S., & Anwar, S., (2018). *GPS Guided Autonomous Navigation of a Small Agricultural Robot with Automated Fertilizing System*. SAE Tech. Paper, 2018-01-0031.

28. Suradhaniwar, S., Kar, S., Nandan, R., Raj, R., & Jagarlapudi, A., (2018). Geo-ICDTs: Principles and applications in agriculture. In: Reddy, G., & Singh, S., (eds.), *Geospatial Technologies in Land Resources Mapping, Monitoring, and Management* (Vol. 21). Cham, Switzerland: Springer.

29. Kim, S., Lee, M., & Shin, C., (2018). IoT-based strawberry disease prediction system for smart farming. *Sensors, 18*, 4051.

30. Venkatesan, R., Kathrine, G., Jaspher, W., & Ramalakshmi, K., (2018). Internet of Things based pest management using natural pesticides for small scale organic gardens, *J. Comput. Theor. Nanosci., 15*, 2742–2747.

31. James, W., (2014). *Remote Pest Management with Automated Traps*. Available: https://blog.semios.com/remote-pest-management-with-automated-traps (accessed on 16 November 2021).

32. Spensa Z-Trap. Available: http://spensatech.com/ (accessed on 16 November 2021).

33. Wietzke, A., Westphal, C., Gras, P., Kraft, M., Pfohl, K., Karlovsky, P., Pawelzik, E., et al., (2018). Insect pollination is a key factor for strawberry physiology and marketable fruit quality. *Agricult., Ecosyst. Environ., 258*, 197–204.

34. Chung, S. O., Choi, M. C., Lee, K. H., Kim, Y. J., Hong, S. J., & Li, M., (2016). Sensing technologies for grain crop yield monitoring systems: A review. *J. Biosyst. Eng., 41*, 408417.

35. https://www.farmtrx.com/ (accessed on 16 November 2021).

36. Udomkun, P., Nagle, M., Argyropoulos, D., Mahayothee, B., & Müller, J., (2016). Multi-sensor approach to improve optical monitoring of papaya shrinkage during drying, *J. Food Eng., 189*, 82–89.

37. http://mx.nthu.edu.tw/~yucsu/3271/p08.pdf (accessed on 16 November 2021).

38. Gasso-Tortajada, V., Ward, A. J., Mansur, H., Brøchner, T., Sørensen, C. G., & Green, O., (2010). A novel acoustic sensor approach to classify seeds based on sound absorption spectra. *Sensors, 10*, 10027–10039.

39. Srivastava, N., Chopra, G., Jain, P., & Khatter, B., (2013). Pest monitor and control system using wireless sensor network (with special reference to acoustic device wireless sensor). *Proc. 27*[th] *Int. Conf. Elect. Electron. Eng.* (pp. 1–7). Goa, India.

40. https://www.elprocus.com/optical-sensors-types-basics-and-applications/ (accessed on 16 November 2021).

41. Murray, S. C., (2018). Optical sensors advancing precision in agricultural production. *Photon. Spectra, 51*, 48.

42. Povh, F. P., Paula, G. D. A. W. D., (2014). Optical sensors applied in agricultural crops. *Opt. Sensors-New Develop. Practical Appl.*

43. Pajares, G., (2011). Advances in sensors applied to agriculture and forestry. *Sensors, 11*, 89308932, 2011.

44. https://www.forschungsfabrik-mikroelektronik.de/en/Range_Of_Services/Technologies/Sensor_Systems.html (accessed on 16 November 2021).

45. Andújar, D., Ribeiro, R., Fernández-Quintanilla, C., & Dorado, J., (2011). Accuracy and feasibility of optoelectronic sensors for weed mapping in wide row crops. *Sensors, 11*, 2304–2318.

46. Andãºjar, D., Ribeiro, A., Quintanilla, C. F., Dorado, J., & Dorado, J., (2009). Assessment of a ground-based weed mapping system in maize. *Precision Agriculture*.

47. https://www.fierceelectronics.com/sensors/what-ultrasonic-sensor (accessed on 16 November 2021).

48. Dvorak, J. S., Stone, M. L., & Self, K. P., (2016). Object detection for agricultural and construction environments using an ultrasonic sensor. *J. Agricult. Saf. Health, 22*, 107–119.

49. Álvarez-Arenas, T. G., Gil-Pelegrin, E., Cuello, J. E., Fariñas, M. D., Sancho-Knapik, D., Burbano, D. A. C., & Peguero-Pina, J. J., (2016). Ultrasonic sensing of plant water needs for agriculture. *Sensors, 16*, 1089.

50. https://www.electronics-tutorials.ws/io/io_3.html (accessed on 16 November 2021).

51. Chazette, L., Becker, M., & Szczerbicka, H., (2016). Basic algorithms for beehive monitoring and laser-based mite control. In: *Proceedings of the 2016 IEEE Symposium Series on Computational Intelligence (SSCI)* (pp. 1–8). Athens, Greece.

52. García-Ramos, F. J., Vidal, M., Boné, A., Malón, H., & Aguirre, J., (2012). Analysis of the airflow generated by an air-assisted sprayer equipped with two axial fans using a 3D sonic anemometer. *Sensors, 12*, 7598–7613.

53. Jason, N. S., Matthew, J. D., & Robert, P. M., (2017). *Performance Benchmark of Yield Monitors for Mechanical and Environmental Influences*. Iowa State University Digital Repository. https://lib.dr.iastate.edu/cgi/viewcontent.cgi?article=1522&context=abe_eng_conf.

54. https://www.mouser.in/applications/smart-agriculture-sensors/ (accessed on 16 November 2021).

55. https://www.analyticaltechnology.com/analyticaltechnology/gas-water-monitors/blog.aspx?ID=1327&Title=How%20Do%20Electrochemical%20Sensors%20Work/ (accessed on 16 November 2021).

56. Yew, T. K., Yusoff, Y., Sieng, L. K., Lah, H. C., Majid, H., & Shelida, N., (2014). An electrochemical sensor ASIC for agriculture applications. *Proc. 37th Int. Conv. Inf. Commun. Technol., Electron. Microelectron. (MIPRO)*. Opatija, Croatia, 8590.

57. https://www.enas.fraunhofer.de/en/business_units/sensor_and_actuator_systems/Electromagnetic_sensors.html (accessed on 16 November 2021).

58. https://blog.agrivi.com/post/smart-sensors-for-accurate-soil-measurements (accessed on 16 November 2021).

59. Hemmat, A., Binandeh, A. R., Ghaisari, J., & Khorsandi, A., (2013). Development and field testing of an integrated sensor for on-the-go measurement of soil mechanical resistance. *Sens. Actuators A, Phys., 198*, 61–68.

60. Kumar, A., Bhatia, A., Fagodiya, R. K., Malyan, S. K., & Meena, B. L., (2017). Eddy covariance flux tower: A promising technique for greenhouse gases measurement. *Adv. Plants Agricult. Res., 7*, 337–340.

61. Del-Moral-Martínez, I., Rosell-Polo, J. R., Company, J., Sanz, R., Escolà, A., Masip, J., Martínez-Casasnovas, J. A., & Arnó, J., (2016). Mapping vineyard leaf area using mobile terrestrial laser scanners: Should rows be scanned on the go or discontinuously sampled? *Sensors, 16*, 119.

62. Biber, P., Weiss, U., Dorna, M., & Albert, A., (2012). *Navigation System of the Autonomous Agricultural Robot-'BoniRob.'* http://www.cs.cmu.edu/~mbergerm/agrobotics2012/01Biber.pdf (accessed on 16 November 2021).

63. Tiwari, D. K., Dasgupta-Schubert, N., Villasen, C. L. M., Villegas, J., Carreto, M. L. S., & Borjas, G. E., (2014). *Interfacing Carbon Nanotubes (CNT) with Plants: Enhancement of Growth, Water, and Ionic Nutrient Uptake in Maize (Zea mays) and Implications for Nanoagriculture, 4*, 577–591. doi: 10.1007/s13204-013-0236-7.

64. Srivastava, A., & Rao, D. P., (2014). Enhancement of seed germination and plant growth of wheat, maize, peanut, and garlic using multiwalled carbon nanotubes. *Eur. Chem. Bull., 3*, 502–504.

65. Husenand, A., & Salahuddin, S. K., (2014). Carbon and fullerene nanomaterials in plant system. *Journal of Nanobiotechnology, 12*. doi:10.1186/1477-3155-12-16.

66. Tai-Chia, C., & Chih-Ching, H., (2009). Aptamer-functionalized nano-bio sensors. *Sensors.* ISSN 1424, 8220, 10356–10388.

67. McKeague, M., Giamberardino, A., & DeRosa, M. C., (2011). Advances in aptamer-based biosensors for food safety. *Environmental Biosensors.* ISBN: 978-953-307-4863.

68. Bawankar, S. D., Bhople, S. B., & Jaiswal, V. D., (2012). Mobile networking for "smart dust" with RFID sensor networks. *International Journal of Smart Sensors and Ad Hoc Networks.* ISSN No. 2248–9738, 2.

69. McGonigle, A. J. S., Wilkes, T. C., Pering, T. D., Willmott, J. R., Cook, J. M., Mims, F. M., & Parisi, A. V., (2018). Smartphone spectrometers. *Sensors, 18*, 223.

70. Camacho, A., & Arguello, H., (2018). Smartphone-based application for agricultural remote technical assistance and estimation of visible vegetation index to farmer in Colombia: AgroTIC. *Proc. SPIE*, 10783, 107830K.

71. Han, P., Dong, D., Zhao, X., Jiao, L., & Lang, Y., (2016). A smartphone-based soil color sensor: For soil type classification. *Comput. Electron. Agricult., 123*, 232241.

72. Xie, X., Zhang, X., He, B., Liang, D., Zhang, D., & Huang, L., (2016). A system for diagnosis of wheat leaf diseases based on Android smartphone. *Proc. SPIE*, 10155.

73. Stiglitz, R., Mikhailova, E., Post, C., Schlautman, M., Sharp, J., Pargas, R., Glover, B., & Mooney, J., (2017). Soil color sensor data collection using a GPS-enabled smartphone application. *Geoderma, 296*, 108–114.

74. Frommberger, L., Schmid, F., & Cai, C., (2013). Micro-mapping with smartphones for monitoring agricultural development. *Proc. 3rd ACM Symp. Comput. Develop., 46.*

75. Kou, Z., & Wu, C., (2018). Smartphone based operating behavior modeling of agricultural machinery. *IFAC-PapersOnLine, 51*, 521–525.

76. Wan, X., Cui, J., Jiang, X., Zhang, J., Yang, Y., & Zheng, T., (2018). Smartphone based hemispherical photography for canopy structure measurement. *Proc. SPIE*, 10621, 106210Q.

77. Debauche, O., Mahmoudi, S., Andriamandroso, A. L. H., Manneback, P., Bindelle, J., & Lebeau, F., (2018). Cloud services integration for farm animals' behavior studies based on smartphones as activity sensors. *J. Ambient Intell. Humaniz. Comput.*, 1–12.

78. Azam, M. F. M., Rosman, S. H., Mustaffa, M., Mullisi, S. M. S., Wahy, H., Jusoh, M. H., & Ali, M. I., (2016). Hybrid water pump system for hilly agricultural site. *Proc. 7th IEEE Control Syst. Graduate Res. Colloq. (ICSGRC)*, 109–114.

79. How to Feed the World in 2050 by FAO. https://www.fao.org/wsfs/forum2050/wsfs-forum/en/ (accessed on 16 November 2021).

80. Mark, T., & Griffin, T., (2016). Defining the barriers to telematics for precision agriculture: Connectivity supply and demand. *SAEA Annu. Meeting.* Austin, TX, USA.

81. Mohamed, A. K. E., (2013). *Analysis of Telematics Systems in Agriculture.* M.S. thesis, Dept. Mach., Utilization, CULS, Prague, Czech Republic.

82. Patmasari, R., Wijayanto, I., Deanto, R. S., Gautama, Y. P., & Vidyaningtyas, H., (2018). Design and realization of automatic packet reporting system (APRS) for sending telemetry data in Nanosatellite communication system. *J. Meas., Electron., Commun., Syst., 4,* 1–7.

83. Turner, J. D., & Austin, L., (2000). A review of current sensor technologies and applications within automotive and traffic control systems. *Proceedings of the Institution of Mechanical Engineers, Part D: Journal of Automobile Engineering, 214,* 589–614. https://doi.org/10.1243/0954407001527475.

84. Vågen, T. G., Winowiecki, L. A., Tondoh, J. E., Desta, L. T., & Gumbricht, T., (2016). Mapping of soil properties and land degradation risk in Africa using MODIS reactance. *Geoderma, 263,* 216–225.

85. Jaafar, H. H., & Woertz, E., (2016). Agriculture as a funding source of ISIS: A GIS and remote sensing analysis. *Food Policy, 64,* 14–25.

86. Rose, I., & Welsh, M., (2010). Mapping the urban wireless landscape with Argos. *Proc. 8th ACM Conf. Embedded Netw. Sensor Syst. (SenSys)* (pp. 323–336). New York, NY, USA.

87. Zhu, L., Suomalainen, J., Liu, J., Hyyppä, J., Kaartinen, H., & Haggren, H., (2018). A review: Remote sensing sensors. In: Rustamov, R. B., Hasanova, S., & Zynalova, M. H., (eds.), *Multi-purposeful Application of Geospatial Data.* London, UK: IntechOpen.

88. Saeed, I. A., Wang, M., Ren, Y., Shi, Q., Malik, M. H., Tao, S., Cai, Q., & Gao, W., (2019). Performance analysis of dielectric soil moisture sensor. *Soil and Water Research, Short Communication, 14,* 195–199.

89. De La Piedra, A., Braeken, A., & Touha, A., (2012). Sensor systems based on FPGAs and their applications: A survey. *Sensors, 12,* 12235–12264.

90. Husni, M. I., Hussein, M. K., Zainal, M. S. B., Hamzah, A., Nor, D. M., & Poad, H., (2018). Soil moisture monitoring using eld programmable gate array, *Indonesian J. Elect. Eng. Comput. Sci., 11,* 169174.

91. Millan-Almaraz, J. R., De Romero-Troncoso, R. J., Guevara-Gonzalez, R. G., Contreras-Medina, L. M., Carrillo-Serrano, R. V., Osornio-Rios, R. A., Duarte-Galvan, C., et al., (2010). FPGA-based fused smart sensor for real-time plant-transpiration dynamic estimation. *Sensors, 10,* 8316–8331.

92. Husni, M. I., Hussein, M. K., Zainal, M. S. B., Hamzah, A., Nor, D. M., & Poad, H., (2018). Soil moisture monitoring using field programmable gate array. *Indonesian J. Elect. Eng. Comput. Sci., 11,* 169–174.

93. Molina, I., Morillo, C., García-Meléndez, E., Guadalupe, R., & Roman, M. I., (2011). Characterizing olive grove canopies by means of ground based hemispherical photography and spaceborne RADAR data. *Sensors, 11,* 74767501.

94. Santhi, P. V., Kapileswar, N., Chenchela, V. K. R., & Prasad, C. H. V. S., (2017). Sensor and vision based autonomous AGRIBOT for sowing seeds. *Proc. Int. Conf. Energy, Commun., Data Anal. Soft Comput.* Chennai, Tamil Nadu, 242245.

95. Karimi, H., Navid, H., Besharati, B., Behfar, H., & Eskandari, I., (2017). A practical approach to comparative design of non-contact sensing techniques for seed ow rate detection, *Comput. Electron. Agricult., 142,* 165172.

96. https://www.rcrwireless.com/20161005/Internet of Things/precision-agriculture-bean-tag31-tag99 (accessed on 16 November 2021).

97. Jeziorska, J., (2019). UAS for wetland mapping and hydrological modeling. *Remote Sensing, 11.* https://doi.org/10.3390/rs11171997.

98. Zujevs, A., Osadcuks, V., & Ahrendt, P., (2015). Trends in robotic sensor technologies for fruit harvesting: 2010–2015. *Procedia Comput. Sci., 77,* 227–233.

99. Zhu, C., Leung, V., Shu, L., & Ngai, E. C. H., (2015). Green Internet of Things for smart world. *Access IEEE, 3,* 2151–2162.

100. Shaikh, Faisal, K., et al., (2017). Enabling technologies for green Internet of Things. *IEEE Systems Journal, 11,* 983–994.

101. Anastasi, G., Conti, M., Francesco, M. D., & Passarella, A., (2009). Energy conservation in wireless sensor networks: A survey. *Ad Hoc Netw., 7,* 537–568.

102. Rault, T., Bouabdallah, A., & Challal, Y., (2014). Energy efficiency in wireless sensor networks: A top-down survey, *Comput. Netw., 67,* 104–122.

103. Ojha, T., Bera, S., Misra, S., & Raghuwanshi, N. S., (2014). Dynamic duty scheduling for green sensor-cloud applications. *Proceedings of IEEE CloudCom.* Singapore.

104. Zhu, C., Yang, L. T., Shu, L., Rodrigues, J. J. P. C., & Hara, T., (2012). A geographic routing oriented sleep scheduling algorithm in duty-cycled sensor networks. *Proc. IEEE Int. Conf. Commun.,* 5473–5477.

105. Tawk, Y., Costantine, J., & Christodoulou, C. G., (2014). Cognitive-radio and antenna functionalities: A tutorial [wireless corner]. *IEEE Antennas Propag. Mag., 56,* 231–243.

106. Shu, L., Zhang, Y., Yang, L. T., Wang, Y., Hauswirth, M., & Xiong, N., (2010). TPGF: Geographic routing in wireless multimedia sensor networks. *Telecommun. Syst., 44,* 79–95.

107. Zhu, C., Yang, L. T., Shu, L., Leung, V. C. M., Rodrigues, J. J. P. C., & Wang, L., (2014). Sleep scheduling for geographic routing in duty-cycled mobile sensor networks. *IEEE Trans. Ind. Electron., 61,* 6346–6355.

108. Tombaz, S., Vastberg, A., & Zander, J., (2011). Energy- and cost-efficient ultra-high-capacity wireless access. *IEEE Wireless Commun., 18,* 18–24.

109. Karakus, C., Gurbuz, A. C., & Tavli, B., (2013). Analysis of energy efficiency of compressive sensing in wireless sensor networks. *IEEE Sensors J., 13,* 1999–2008.

110. Sheng, Z., Fan, J., Liu, C. H., Leung, V. C. M., Liu, X., & Leung, K. K., (2015). Energy-efficient relay selection for cooperative relaying in wireless multimedia networks. *IEEE Trans. Veh. Technol., 64,* 1156–1170.

111. Ruan, J., Wang, Y., Chan, F. T. S., Hu, X., Zhao, M., Zhu, F., Shi, B., Shi, Y., & Lin, F., (2019). A life cycle framework of green iot-based agriculture and its finance, operation, and management issues. *IEEE Commun. Mag., 57,* 90–96.

112. http://loupelectronics.com/products/yield_monitor.html (accessed on 16 November 2021).

113. http://www.icstation.com/m214-soil-moisture-sensor-humidity-controller-module-99rh-automatic-control-irrigation-system-digital-display-controller-p-13099.html (accessed on 16 November 2021).

114. https://agriculture.trimble.com/product/ag-premium-weather/ (accessed on 16 November 2021).
115. http://phyto-sensor.com/FILM-FI-MM-FI-SM (accessed on 16 November 2021).
116. http://www.ictinternational.com/products/mp406/mp406-moisture-sensor/ (accessed on 16 November 2021).
117. https://www.deere.com/en/technology-products/precision-ag-technology/guidance/greenstar-3-2630/ (accessed on 16 November 2021).
118. Ediz, H., Florian, S., Nikolai, S., Ramon, W., & Harald, G., (2020). *Low-Cost Hydrogen Sensor in the ppm Range with Purely Optical Readout ACS Sens.*, *5*(4), 978–983.
119. http://www.dynamax.com/products/plant-growth-sensors/dex-fruit-stem-growth-dendrometer (accessed on 16 November 2021).
120. Troy, S. M., Christian, F., Joshua, B. F., Ying, S., Gretchen, B. N., Thomas, S. D., Ari, K., & Katharina, S., (2017). *Connecting Active to Passive Fluorescence with Photosynthesis: A Method for Evaluating Remote Sensing Measurements of Chl Fluorescence* (Vol. 215, No. 4, pp. 1594–1608).
121. https://www.campbellsci.com/03002-wind-sentry (accessed on 16 November 2021).
122. https://www.aeroqual.com/aqm-65-air-monitoring-station (accessed on 16 November 2021).
123. https://www.metergroup.com/environment/articles/which-soil-sensor-is-perfect-for-you/ (accessed on 16 November 2021).
124. Davis, T. W., Kuo, C. M., Liang, X., & Yu, P. S., (2012). Sap flow sensors: Construction, quality control and comparison. *Sensors, 12*(1), 954–971. https://doi.org/10.3390/s120100954.
125. Mestre, G., Ruano, A., Duarte, H., Silva, S., Khosravani, H., Pesteh, S., Ferreira, P. M., & Horta, R., (2015). An intelligent weather station. *Sensors*, *15*(12), 31005–31022. https://doi.org/10.3390/s151229841.
126. https://citizensense.net/kits/phyto-sensor-toolkit/ (accessed on 16 November 2021).
127. https://www.farms.com/precision-agriculture/yield-monitoring/ (accessed on 16 November 2021).
128. Bitponics, http://www.bitponics.com/ (accessed on 16 November 2021).
129. Botanicalls, http://www.botanicalls.com/ (accessed on 16 November 2021).
130. Edyn, http://www.saturnbioponics.com/ (accessed on 16 November 2021).
131. Parrot: Flower Power, http://www.parrot.com/ (accessed on 16 November 2021).
132. Plantlink, http://myplantlink.com/ (accessed on 16 November 2021).
133. HarvestGeek, http://www.harvestgeek.com/ (accessed on 16 November 2021).
134. Wi-Fi Sprinkler System: Iro, https://rachio.com/ (accessed on 16 November 2021).
135. Spruce, https://spruceirrigation.com/sensors/ (accessed on 16 November 2021).
136. Koubachi, https://www.techhive.com/article/3220478/best-smart-sprinkler-controller.html (accessed on 16 November 2021).
137. Niwa, www.getniwa.com (accessed on 16 November 2021).
138. Why IoT, (2019). *Big Data & Smart Farming are the Future of Agriculture*. Available: https://www.businessinsider.com/Internet of Things-smart-agriculture-2016-10 (accessed on 16 November 2021).
139. Ojha, T., Misra, S., & Raghuwanshi, N. S., (2013). Wireless sensor networks for agriculture: The state-of-the-art in practice and future challenges. *Computers and Electronics in Agriculture, 118*, 66–84. doi: 10.1016/j.compag.2013.08.011.

140. Zhu, C., Leung, V., Shu, L., & Ngai, E. C. H., (2015). Green Internet of Things for smart world. *Access, IEEE, 3*, 2151–2162.
141. https://www.mouser.in/applications/smart-agriculture-sensors/ (accessed on 16 November 2021).
142. Rupanagudi, S. R., Ranjani, B. S., Nagraj, P., Bhat, V. G., & Thippeswamy, G., (2015). A novel cloud computing based smart farming system for early detection of borer insects in tomatoes. *Proceedings of International Conference on Communication, Information & Computing Technology (ICCICT)*. Mumbai, India.
143. https://www.agritechtomorrow.com/article/2019/02/top-article-for-2019-smart-sensors-in-farming/11247 (accessed on 16 November 2021).
144. Sharma, D., Shukla, A. K., Bhondekar, A. P., Ghanshyam, C., & Ojha, A., (2016). A Technical assessment of IoT for Indian agriculture sector. *IJCA Proceedings on National Symposium on Modern Information and Communication Technologies for Digital India MICTDI, 1*, 1–5.
145. Ayaz, M., Ammad-Uddin, M., Sharif, Z., Mansour, A., & Aggoune, E. M., (2019). Internet-of-things (IoT)-based smart agriculture: Toward making the fields talk. *IEEE Access, 7*, 129551–129583. doi: 10.1109/ACCESS.2019.2932609.

Index